Sustainable Energy Access for Communities

Aminata Fall • Reinhard Haas
Editors

Sustainable Energy Access for Communities

Rethinking the Energy Agenda for Cities

Springer

Editors
Aminata Fall
Energy Economics Group
TU Wien
Vienna, Austria

Reinhard Haas
Energy Economics Group
TU Wien
Wien, Austria

ISBN 978-3-030-68409-9 ISBN 978-3-030-68410-5 (eBook)
https://doi.org/10.1007/978-3-030-68410-5

This Springer imprint is published by the registered company Springer Nature Switzerland AG.
The registered company address is: Gewerbestrasse 11, 6330 Cham, Switzerland

Foreword

Strengthening Access to Sustainable Energy for All

With Sustainable Development Goal (SDG) number 7—ensuring access to affordable, reliable, sustainable, and modern energy for all—the United Nations' 2030 Agenda emphasises the significance of energy access. According to the International Energy Agency's most recent data, the number of people without access to electricity had been decreasing since 2013. However, because of the COVID-19 pandemic, this number is set to rise again in 2020, threatening past efforts and achievements to make energy accessible to everyone. Nonetheless, efforts to make energy accessible to all should not be at any cost. Rising sea levels, droughts, floods, and intensifying natural disasters have taught us that we cannot rely on fossil fuels in the pathway to a sustainable energy future. Over the next few years, adapting to climate change and mitigating its disastrous effects on the most vulnerable populations will remain the international community's biggest challenge. Even though global energy demand declined in 2020, reducing carbon emissions, the global community is not yet on the right path to achieve the net-zero target by 2050. The importance of the nexus between access to clean energy and sustainable development cannot be overstated; it is visible in cities and rural areas, in rich and poor countries, in research and in practice.

The Austrian Development Agency (ADA), the operational unit of Austrian Development Cooperation, recognises the instrumental role of access to clean energy in a better future for everyone. Together with the United Nations Industrial Development Organization (UNIDO), ADA has been supporting the creation and expansion of a global network of regional centres for renewable energy and energy efficiency for about 10 years. In addition to financing knowledge hubs such as this initiative, further funding is earmarked for research on energy-related thematic, as well as for connecting experts and experts-to-be. In this context, the Austrian

Development Cooperation's APPEAR[1] programme plays a crucial role. Amongst other key themes, APPEAR funds research on sustainable energy in and for participating countries, which provided the starting point for the research findings presented in this book.

We, as ADA, are welcoming the publication *Sustainable Energy Access for Sustainable Communities: Rethinking the Energy Agenda for Cities.* This book brings together researchers and professionals from across the field, provides answers to striking questions of our time, and makes an important contribution to reaching SDG-7.

On behalf of the Austrian Development Agency and APPEAR, I would like to thank the authors who contributed to this book for putting sustainability at the forefront of the research agenda, and for their engagement in documenting a thematic that is at the intersection between international cooperation and university research. I trust that this publication will receive prominent attention, and will prove to be a valuable resource for academic and non-academic stakeholders in our partner countries and beyond.

Austrian Development Agency (ADA), Martin Ledolter
the operational unit of Austrian
Development Cooperation
Vienna, Austria

[1]The Austrian Partnership Programme for Higher Education and Research for Development (APPEAR) supports higher education and research partnerships between institutions in Austria and partner countries of the Austrian Development Cooperation. Since 2010, APPEAR has supported 45 partnerships in 20 countries as well as 126 scholarships, mainly for PhD studies in Austria. The design and implementation of projects is led by the programme's guiding principles: participatory approach, culturally open-minded knowledge, practically and empirically-oriented approach, gender sensitivity, as well as bottom-up and demand-driven approach. For more information, please visit: www.appear.at.

Acknowledgements

The publication acknowledges support of the project Sustainable Energy Access for Sustainable Cities (SEA4cities) that is an academic partnership between Ecole Polytechnique Thies (Senegal) and Vienna University of Technology (Austria), which is funded by the Austrian Partnership Programme in Higher Education and Research for Development (APPEAR).[2]

An academic partnership project of:

[2]APPEAR is a programme of the Austrian Development Cooperation and is implemented by the OeAD.

Supported by:

appear

Austrian Partnership Programme
in Higher Education and Research
for Development

a programme of the

Austrian
Development
Cooperation

About This Book

This contributing volume, *Sustainable Energy Access for Sustainable Communities: Rethinking the Energy Agenda for Cities*, is prepared in the framework of the project Sustainable Energy Access for Sustainable Cities (SEA4cities). The compendium of writings is intended to shed light on the multifaceted dimensions of energy sustainability in different regions covered by the Austrian Partnership Programme in Higher Education and Research for Development (APPEAR). The book aims at bringing at the forefront of the research agenda latest developments on energy access and transition to sustainability in the context of academia and international cooperation. Sustainability in practice demonstrated the need to constitute a large coalition of efforts to support the effectiveness of energy access and systems sustainability throughout the overall value chain: from basic research in universities to documentation of lessons learned in the field. The book relays voices of dedicated researchers and professionals who share not only their work, experiments, and findings, but also their driving motivation and passion for a thematic we define as the challenge of our generation.

Access to sustainable energy sounds like a craze nowadays; everybody knows it is the goal, few, if any, can indicate the path to get there. The reason is the many dimensions of access and transition to sustainability in the energy sector raise complex questions that we are finding increasingly difficult to provide a holistic answer. Here, by complex, we mean the usual binary questions that over the decades have evolved to become interconnected hydra with multifaceted dimensions. Today, when we question the way to energy sustainability in an emerging city, such as Dakar in Senegal[3], we will necessarily address the challenges of economic growth, social stability, education, water supply, waste management, etc. This complexity explains why the answer cannot be restrictive or even be an answer at all. There is no

[3]The word emerging for the city of Dakar considers classification of the World Cities by the New Climate Economy Report (2014), meaning "rapidly expanding middle-income, mid-sized cities (...), with populations of 1–10 million, and per capita incomes of US$2,000-20,000".

single solution to the equation made of these challenges, but there are local optima to reach through collective efforts. This approach is that of the book. The contributions bring together junior researchers with relevant fundamental questions and experienced researchers and professionals in the field of energy. Through these interactions, the authors drafted 15 chapters that open new perspectives in thinking access and transition to energy sustainability as:

- A mechanism for a participatory governance closer to local communities
- A kingpin of technology innovation and driver of a green digital revolution
- A social concern that stems from business-as-usual because the long-term goal of climate mitigation should not divert from the immediate objectives of fulfilling basic needs.

In this sense, the book brings a valuable contribution to contemporary debates. It fits in a momentum of action, where cities whether mature or emerging take measures to make their energy mix greener, reduce their environmental footprints, and improve their living standards. The cities, we present in this book, currently address these challenges through diverse initiatives. Our contributors present some of these initiatives, discuss the lessons learned from their implementation, and propose new avenues for better. As we expect to reach a new optimum on thinking energy access and transition to sustainability, we lay the basis for the current and new generation of researchers to build a step further towards our common goal to ensure universal access to sustainable and modern energy services (SDG-7).

Contents

Contributors

Convening Lead Author (CLA)

Aminata Fall Energy Economics Group, TU Wien, Austria

Contributing Authors (In Order of Chapters)

Martin Ledolter Austrian Development Agency–Foreword, Vienna, Austria

Andreas J. Obrecht Austrian Partnership Programme in Higher Education and Research for Development, Vienna, Austria

Abdoulaye Thiaw Faculty of Economics–UCAD, Municipality of Pikine, Dakar, Senegal

Sidia Badiane Department of Geography – UCAD, Dakar, Senegal

Cheikh Faye Ministry of Labour, Business Integration Programme, Dakar, Senegal

Pauline Macharia Institute for Water Quality and Resource Management, TU Wien, Austria

Cheikh A. Mbodji Research Lab Electronics, Informatics, IT and Renewable Energy, UGB, Saint-Louis, Senegal

Christof Sumereder FH Joanneum, University of Applied Sciences, Kapfenberg, Austria

Djicknoum Diouf Research Lab Electronics, Informatics, IT and Renewable Energy, UGB, Saint-Louis, Senegal

Amadou S. H. Maiga Research Lab Electronics, Informatics, IT and Renewable Energy, UGB, Saint-Louis, Senegal

Reinhard Haas Energy Economics Group, TU Wien, Austria

Abdoulaye Seck SEnergyS Africa Co., Dakar, Senegal

Cheikh Ahmet T. C. Ndiaye Polytechnic School of Thies, Thies, Senegal

Amadou L. Ndour Responsibility Africa Ltd., Nairobi, Kenya

Amadou L. Fall Global Green Growth Institute, Dakar, Senegal

Willy M. Nkounga Polytech Annecy-Chambery, France and Polytechnic School of Dakar, Dakar, Senegal

Mouhamadou F. Ndiaye Polytechnic School of Dakar, UCAD, Dakar, Senegal

Mamadou L. Ndiaye Polytechnic School of Dakar, UCAD, Dakar, Senegal

Saliou Toure United Nations Development Programme, Istanbul, Turkey

Salif Sow Research Lab Water, Energy, Environment and Industrial Processes, Polytechnic School of Dakar, UCAD, Dakar, Senegal

Cheikh M. F. Kebe Research Lab Water, Energy, Environment and Industrial Processes, Polytechnic School of Dakar, UCAD, Dakar, Senegal

Ababacar Ndiaye Research Lab Water, Energy, Environment and Industrial Processes, Polytechnic School of Dakar, UCAD, Dakar, Senegal

Mamadou Wade Polytechnic School of Thies, Thies, Senegal

Review Editors (RE)

Aminata Fall Energy Economics Group, TU Wien, Austria

Reinard Haas Energy Economics Group, TU Wien, Austria

List of Abbreviations

ADA	Austrian Development Agency
APPEAR	Austrian Partnership Programme in Higher Education and Research for Development
CO_2	Carbon dioxide
CO_2e	Greenhouse gas emissions in equivalent CO_2, considering the radiative forcing of the greenhouse gases
EUR	Euro (European currency)
ENPEP–BALANCE	Energy and Power Evaluation Program (modelling software)
GJ	Gigajoules
HDI	Human development index
IEA	International Energy Agency
INDC	Intended Nationally Determined Contribution to climate change mitigation
LEAP	Low Emissions Analysis Platform (modelling software)
MoCES	Modelling Cities Energy Systems
SDG	Sustainable Development Goal
SEA4cities	Sustainable Energy Access for Sustainable Cities
SSA	Sub-Saharan Africa
Toe	Tonne of oil equivalent (an energy metric)
USD	Dollar of the United States
W	Watt; related metrics are kW (kilowatt) and MW (megawatt)

List of Figures

List of Tables

Chapter 1
Sustainable Energy Access for Sustainable Communities: Introduction by a Social Scientist

Andreas J. Obrecht

Abstract Three pillars are essential to building a sustainable future for current and next generations: sustainable energy access, inclusive knowledge and equal opportunities in education, peacebuilding, and democratisation. These pillars interact like communicating vessels; they can neither be conceived nor achieved in isolation from one another. Sustainable energy access relies on access to broad and decentralised knowledge and to innovation in technologies, which is the pursuit of higher education. Community-based implementation of the sustainable energy principles is only possible if the stakeholders own these principles and thus create the basis for participatory, transparent, democratic, and above all, peaceful conditions for action. Peaceful contexts are necessary for communities to achieve the socio-ecological transformation that is necessary to mitigate climate change and to build models of a sustainable coexistence that preserves our common good, which is the environment.

Keywords Sustainable energy access · Inclusive knowledge · Decentralised knowledge

Three pillars are essential to building a sustainable future for current and next generations: sustainable energy access, inclusive knowledge and equal opportunities in education, peacebuilding, and democratisation. These pillars interact like communicating vessels; they can neither be conceived nor achieved in isolation from one another. Sustainable energy access relies on access to broad and decentralised knowledge and to innovation in technologies, which is the pursuit of higher education. Community-based implementation of the sustainable energy principles is only possible if the stakeholders own these principles and thus create the basis for participatory, transparent, democratic, and above all, peaceful conditions for action. Peaceful contexts are necessary for communities to achieve the socio-ecological

A. J. Obrecht (✉)
APPEAR—Austrian Partnership Programme in Higher Education and Research for Development, Vienna, Austria
e-mail: Andreas.Obrecht@oead.at

A. Fall, R. Haas (eds.), *Sustainable Energy Access for Communities*,
https://doi.org/10.1007/978-3-030-68410-5_1

transformation that is necessary to mitigate climate change and to build models of a sustainable coexistence that preserves our common good, which is the environment.

APPEAR—the Austrian Partnership Programme in Higher Education and Research for Development—supports projects for sustainable energy access and efficiency in Kenya, Mozambique, Palestine, and Senegal. This book mainly focuses on the SEA4cities project that is being implemented in Senegal while also taking a comparative look at other countries of APPEAR's intervention. APPEAR is the cooperation programme in higher education of the Austrian Development Cooperation (OeZA), which is implemented by the Austrian Agency for Education and Internationalisation (OeAD). The APPEAR's contributions to higher education cover a broad spectrum, addressing all three essential pillars of a sustainable future across three continents (Africa, Asia, and Europe).

1 Unequal Distribution of Access to Energy Worldwide

Of the approximately 7.8 billion people in the world, about 10% have no access to electricity—four-fifths of these people live in rural areas of developing countries. In the future, the number of people without access to electricity could further increase in the rapidly growing cities of developing countries driven by urbanisation. Approximately 1.3 million barrels of oil, which is equivalent to 1.5% of the world's production, are spent every day on lighting in the poorest households (Energypedia, 2017)[1], often in the form of bottles with wicks, which represent a serious health concern. Two and a half (2.5) billion people continue to cook and heat using energy-intensive technologies such as the three-stone fire with raw biomass, i.e. wood, straw, corn stalks, animal dung, often collected by women and young girls. More than half of these people live in India (825 million) and China (515 million); they represent 83% of the 1.2 Sub-Saharan Africans (IEA, 2020)[2].

In 2019, 52% of the Sub-Saharan Africa's population[3] did not have access to electricity, and the global figures veil disparities between urban and rural areas, where access is about 29% (World Energy Outlook, 2020)[4]. Still, the challenge in cities also requires close attention. Sub-Saharan Africa recorded the highest rate of urbanisation over the last two decades; the United Nations (2018) expects that the number of cities with 500,000 inhabitants or more will grow by 57% between 2018 and 2030. The energy system of most African cities, today, is neither sized nor

[1] Energypedia 2017accessible at: https://energypedia.info/wiki/Main_Page

[2] IEA 2020: Database of the International Energy Agency on Cooking accessible at: https://www.iea.org/reports/sdg7-data-and-projections/access-to-clean-cooking

[3] Sub-Saharan Africa encompasses 49 of the 54 African UN member countries.

[4] World Energy Outlook 2020 accessible at: https://www.iea.org/reports/world-energy-outlook-2020

equipped for meeting the challenges of increasing access and transition to sustainability in the near future.

2 Energy Access for Improving Living Standards

Due to the large disparities in access to energy, especially to electricity, there is a concern that triggering the transition to sustainability may affect the energy access targets; challenges related to investment in infrastructure and intermittency of supply continue to raise concerns on the potential of renewable energy resources to meet the rapid increase of energy demand in sub-Saharan Africa. These concerns assume that the growth in energy demand will continue at the same pace and underestimate the potential of energy efficiency in per capita energy consumption while improving living standards and lifting millions of people out of poverty. In other words, a high standard of living can be achieved and, above all, maintained without following the path of industrialised countries over the last 244 years, since James Watt invented the steam engine. Over the last years, we observed that living standards had been correlated with better efficiency of appliances, and therefore decreasing per capita energy consumption. In the future, people will not be better fed, healthier, better educated, with more energy. This observation is confirmed by the Austrian energy expert and physicist Johannes Schmidl, who looked on the possible correlation between the primary energy consumption per capita and the Human Development Index (HDI) of 140 countries. Figure 1.1 plots this correlation. The HDI is frequently used for measuring the living standards of a country. The metric takes into account, among other parameters, the average life expectancy at birth, the adult literacy rate, the school enrolment rate, and the gross domestic product per capita of the population. All these parameters, generated from national statistics, are reliable parameters for measuring poverty and the status of development.

About two-thirds of the countries in the sub-Saharan Africa region have a primary energy consumption per capita equivalent to less than 40 gigajoules (GJ)[5] per year. These countries also have a relatively low HDI. Of the 47 poorest countries in the world—the so-called least developed countries[6]—33 are in the sub-Saharan Africa region (Fig. 1.1).

When the annual primary energy consumption per capita is between 40 and 110 Gigajoules (GJ), the HDI rises sharply at first, before curbing. The world's average primary energy consumption per capita in 2019 was approximately 75 GJ, which is roughly equivalent to the consumption of Turkey, Ukraine, Argentina, and Bhutan. Above 110 GJ per capita, which is roughly equivalent to the energy

[5] 1 GJ $\approx$ 278 kWh $\approx$ 0,3 MWh

[6] Least developed countries (LDCs) are low-income countries confronted with severe structural impediments to sustainable development. They are highly vulnerable to economic and environmental shocks and have low levels of human assets. www.un.org/development/

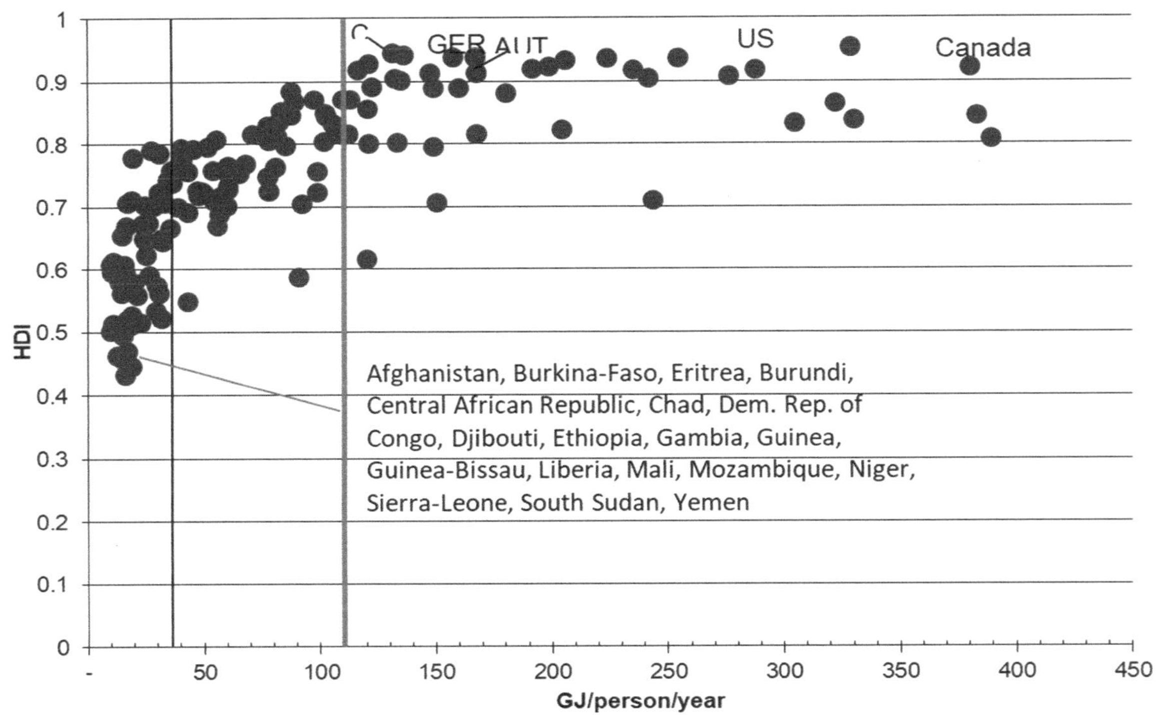

Fig. 1.1 HDI and per capita energy consumption. *Source: Johannes Schmidl (2015): Energie und Utopie. Second Edition. Sonderzahl Publishing House, Vienna*. Fig. 1.1 is based on data of the years 2016 to 2019, depending on availability, and 2018 for the HDI. Primary energy consumption per capita: World Bank, http://data.worldbank.org/indicator; BP Statistical review of World Energy, data on HDI from UNDP: http://hdr.undp.org/en/

consumption of Italy, the United Kingdom and Poland in 2019, there is no statistically discernible gain in quality of life as reflected by the HDI. This figure is the borderline above which additional energy consumption has no influence on the development index and, therefore, on the populations' quality of life. The average energy per capita in the EU was 112 GJ; Austria is in the upper category with 165 GJ per capita. In the USA and Canada, the average energy per capita can be as high as 318 GJ. Countries such as Qatar and Kuwait have averages that exceed these figures by so much that they are outside the above chart. The concept of natural saturation of the primary energy demand, when related to living standards, is good news. It means the process that will lead to the world population peak at around 9.5 billion in 30 years' time, at the latest, will not require energy demand and consumption to rise in proportions similar to those of the last 244 years, in order to improve life standards and lift millions of people out of poverty; this would have been unimaginable a 100 years ago.

Developing countries will continue to increase their primary energy consumption over the next decades in their efforts to align their living standards with the global average. However, the increasing penetration of renewables in electrical grids, from 23% in 2015 to 26.5% according to IEA shows that some net-zero carbon technologies are mature enough to supply this demand. Among the objectives of the Sustainable Development Goal number 7 (SDG 7) is the objective to reduce the unequal distribution of access to electricity—not through gigantomania, large-scale projects that require centralised structures vulnerable to many topical threats, but through smaller and decentralised systems that rely on knowledge- and evidence-based resource management in a participatory and community-based approach. This approach can contribute to achieve by 2030 the socio-ecological transformations needed towards energy sustainability in developing countries.

3 APPEAR Supports academia's Research on Energy Sustainability

The 17 SDGs are the result of decades of scientific research and political negotiations. Fundamental research demonstrated the social and ecological vulnerability of our planet, and science developed practice-oriented solutions to address these "collective action problems." SDG 7 formulates the ambitious goal to "*ensure access to affordable, reliable, sustainable and modern energy for all.*" This goal's attainment primarily requires to progressively phase-out the combustion of conventional energy resources—starting with coal, oil, and natural gas—which are responsible of about 78% of the anthropogenic component of greenhouse gas emissions.[7] An alternative can be the use of locally available renewable energy resources that

[7]The Intergovernmental Panel on Climate Change— IPCC-report 2018, Summary for Policymakers, p. 5: https://www.ipcc.ch/site/assets/uploads/2018/02/AR5_SYR_FINAL_SPM.pdf

provides, in addition to climate change mitigation, an alleviation of the economic system's dependency to energy imports. Local communities in rural and urban areas should be encouraged to become both energy producers and energy consumers with innovative and sustainable technologies.

In order to achieve this goal, which is also highly relevant in development strategies, not only cooperation across countries, institutions, and communities is necessary, but also the contribution of academia is critical. Cooperation with higher education institutions, non-university research institutions, local authorities and government agencies in the partner countries, NGOs and civil society is necessary to build innovative models of energy access and consumption with locally available resources. Innovation in access by means of sustainable energy resources such as solar photovoltaic for electricity generation or waste recycling for process heating is an opportunity to invent models of development that preserve the ecosystems. Especially in Sub-Saharan Africa, it is important that dedicated researchers work with decision-makers on the achievement of the ambitious energy objectives set by national governments and committed to the UN Sustainable Development Goals, in particular the goal number 7.

This research-action approach perfectly tallies with the overarching principles of the APPEAR programme.[8] APPEAR sees itself as a mechanism that supports the achievement of development objectives, of which is the objective to improve scientific capacities in the partner countries and to contribute with knowledge-sharing to improve the social and economic living conditions of the people in these countries. The basis of this ambitious commitment is research, teaching, and innovation that sustain a comprehensive socio-ecological transformation. It is also important that our action today does in no way restrict or even make impossible the life of future generations. It is not possible to think solutions for addressing the environmental problems, especially with regard to climate change, without solutions to address social problems, nor will it be possible to implement the former successfully without the latter. More than ever before, the world needs integrated concepts that approach social and ecological dimensions of topical problems together and develop solutions that simultaneously address them.

APPEAR, the Austrian Development Cooperation's programme in higher education, abide by these guiding principles. We supported the implementation of 45 academic partnerships in 18 partner countries, where our fellow scholars demonstrated the relevance of research and science in bringing solutions to topical problems, which include access and transition to energy sustainability in developing countries. Therefore, we welcome the decision of the Austrian Development Cooperation to continue the APPEAR programme that would have expired at the end of 2020 for another cycle of 7 years. Long-term planning is essential to a solution-oriented cooperation in higher education for development. Time is a critical parameter in forging and sustaining cooperation in science, and to measuring impact on societies.

[8] See www.appear.at

SEA4cities Promotes Research and Cooperation for Energy Sustainability
The APPEAR project SEA4citie addresses, beside technology, the social and socio-political dimensions of access to sustainable energy resources. The project is an institutional cooperation between the Ecole Polytechnique Thies in Senegal and the Vienna University of Technology and aims to create innovative models and tools for transition to energy sustainability in cities. SEA4cities collaborates with other academia and with non-academic partners in both private and public sectors in disseminating tools such as the Energy System Planning Model that contributes to democratise access to energy solutions. The SEA4cities, theoretical concepts, basic research, and field research are methodically linked and presented against the background of the increasing energy demand in sub-Saran Africa. Despite the very different social and economic situations in Europe and sub-Saharan Africa, the challenges of overcoming the dramatic climate change and its consequences on sea-level rise and loss of biodiversity are similar. Thus, the SEA4cities project was able to adopt an integrated perspective from the beginning, building on the experience drawn from initiatives for energy sustainability in Senegal and in Austria. Community grids with renewable energy resources, for example recently gained attention in Europe, particularly in Germany and in Austria. In these countries, academia has largely contributed to defining the models that organise the systems and their operation. SEA4cities supports knowledge exchange between universities in Senegal and Austria to empower a new generation of scientists ready for this challenge.

The book *Sustainable Energy Access for Sustainable Communities: Rethinking the Energy Agenda for Cities*, which compiles findings of a 30-month research-action invites us to embark on a unique journey that explores the multi-faceted dimensions of energy sustainability through its nexus with technology, governance, and development. The book relays views about the meaning of energy sustainability from the perspectives of the project's fellow researchers supported by the contributions of experienced professionals. The meeting of basic research and experience from field practice makes it a valuable contribution to contemporary debates on the topic of energy sustainability, and in doing so, on local participatory governance (Chaps. 2–4). Chapters 5–10 outline the importance of technology innovation and the role of research in building a sustainable energy future. Chapters 11–13 revisit the principle of "common but differentiated responsibilities" by applying it to the level of communities. Chapters 14–15 open perspectives on new pathways for simultaneously achieving the Sustainable Development Goal number 7, and goals on poverty eradication (SDG-1), and access to safe food (SDG-2). The gender dimension in energy sustainability that is addressed in Chap. 13 is of particular importance for APPEAR, because gender equality is a prerequisite for comprehensive participation and is a critical aspect of sustainability in communities.

I thank the convening author, Aminata Fall, for bringing together in this publication exciting contributions from some of the representatives of this new generation of scientists. The book is not only a résumé of interesting research cooperation but also a stimulus for shaping further the path towards a sustainable energy future.

Chapter 2
Local Dimensions of Sustainable Energy Governance: Case Study

Abdoulaye Thiaw and Sidia Badiane

Abstract Innovative scientific approaches and digital tools can support the planning of a sustainable energy transition that complies with the political, economic, social and cultural dimensions of sustainability in local communities. This chapter discusses the leading role of localized energy governance through case studies of projects implemented in Senegal communities. It aims to document the diversity of approaches and tools that are locally available to define and achieve sustainability objectives in the energy sector and thereby contributes to the more inclusive goal of achieving sustainable development. Our ambition is to compare cases with a range of practices that can support municipalities in developing a tailor-made sustainable energy agenda.

Keyword Sustainable Energy · Governance · Communities

1 Introduction

The concept of governance carries a broad shade of meaning. It can refer to the administrative processes and actions initiated by a community of people to respond to their own needs within the framework of a political agenda. Based on this definition, local energy governance can be inferred to mean a combination of measures taken and actions implemented in order to address the populations' needs to access secure, low-carbon, and to the extent possible, affordable energy by leveraging all resources accessible to the municipality. In the energy sector, like in others, effective governance is predicated on inclusiveness, democracy and community buy-in. As Godinot (2011) argues, meeting energy challenges requires a public service mission.

A. Thiaw (✉)
Universite Cheikh Anta Diop Dakar, Dakar, Senegal

S. Badiane
Department of Geography – UCAD, Dakar, Senegal

A. Fall, R. Haas (eds.), *Sustainable Energy Access for Communities*,
https://doi.org/10.1007/978-3-030-68410-5_2

Unlike many other sectors, energy is cross-cutting in nature, as it forms nexus with all other sectors. However, there are numerous barriers impeding the governance of the energy sector in local communities. More than often, energy generation, transmission and distribution powers are the preserve of central governments, and the literature on good practices is recent; major publications on local energy governance were issued as community grids emerged in countries like Germany and Austria. Our study presents these models of energy for communities, including in urban environments, in order to capture the essence of what could be the way forward for other countries. Until now, the literature on decentralized grids in developing countries has focused on rural, remote communities and has been mainly project-based.

The chapter documents the whys and wherefores of governance in Senegalese communities of locally available renewable energy resources. This literature provides us with core practices that can be harnessed to support the transition to energy sustainability in local communities across sub-Saharan Africa. Furthermore, the diversity of environmental, economic and social conditions of communities exemplified should shape responsive programmes of action designed to spur the implementation by local communities of national and international commitments on energy sustainability, such as the Sustainable Development Goal number 7 (SDG-7).

2 Methodological Approach

The approach used to prepare this chapter is a systematic literature review to document the case studies and to support our argument of the existence of basic practices for implementing the energy transition agenda in local communities.

Jon Pierre (2014) argue that "good" local governance should reflect the dualism of efficiency and legitimacy. Local authorities are efficient when they act to provide services that are responsive to local needs and conditions. Local authorities are legitimate when they act as a local branch of the nation-state administrative apparatus. This dual motivation is not conflicting, especially in the energy sector, where it could be reinforcing.

In our approach, we tried to reconcile knowledge from legitimate actions documented in policy strategies such as the *Lettre de Politique de Developpement du Secteur de l'Energie* (LPDSE) in Senegal, which is the reference document of public action in the energy sector for the entire country, and knowledge that derives from learning in doing through local projects and programmes. Literature on central government strategies is publicly accessible online. When preparing this chapter, we were able to access documents that detail the change in the governance framework of the energy sector throughout the last decades. We consulted the literature on projects and initiatives undertaken in local communities, which had an impact on the transition to energy sustainability in order to build the bridge of legitimate and efficient actions for transition to energy sustainability in communities of Senegal and beyond across sub-Saharan Africa.

Our study starts with a presentation of the political framework legally vested in municipalities to pulling the levers that can affect the energy situation of local communities. Then, we present some of the limitations and barriers preventing municipalities from effectively implementing a sustainable energy agenda. This presentation is followed by an introduction of some approaches and tools that could help overcome the barriers identified and usher in a new path for tailor-made governance of energy systems in local communities.

3 Discussion of Findings

The decentralization of political competencies from national to local governments started in Senegal in 1996. Since then, nine areas of responsibility have been transferred "for total administration by local governments (. . .); the central government acting as a controller a posteriori of the regularity of actions" (Ministere en charge des Collectivites Territoriales, 2013). The competencies transferred include land use and waste management. However, the energy supply remains the preserve of the central government. Land use and waste recycling are both critical in the transition to energy sustainability, which creates an overlap of the competencies of local and national authorities. We propose two policy reforms to address this situation:

- Third-party access to the grid, and
- Legal provision of community grids

The first has been gradually introduced in the Senegal legislation since 2011 when the legislation governing self-production, consumption and evacuation of excess energy generation throughout the intercommoned grid has been adopted. The process took seventeen (17) years until the energy regulatory agency proposed the feed-in-tariffs for the evacuation of excess energy generation from renewables. Yet, the legal framework is very restrictive as it only considers self-production of individual buildings and one-way evacuation of the excess energy generated. Our proposed concept of third-party access to the grid is broader and includes the possibility of energy trades between buildings. Darghouth et al. (2015) identify two issues that arise with the concept of third-party access to the grid: (1) variability of power flow (load changes over time) and (2) two-way power flow (end-user can take from and export to the grid). Addressing these issues in local communities of sub-Saran Africa requires a combination of local legislations and infrastructure investment that specifically target these communities. Third-party access to the grid is one step of a more protracted process to make energy infrastructure accessible to all citizens. This is a democratic reform in addition to being an instrument for local municipalities to intervene in the energy sector while being efficient and legitimate. In Europe, the process began when countries decided to decouple energy production and the management of power transmission grids (European Union, 2009). This reform in sub-Saharan Africa should provide municipalities with authority to grant concessions to public and private organizations registered in the municipality for management of the energy transmission grids, thus rethinking the concept of land management, including rights-of-ways. The reform should provide municipalities with increased revenues through investment and taxes paid by concession holders.

The second reform, which provides for legal powers to build community grids, is a step further in democratizing governance of the energy sector. It empowers municipalities as decision-makers and as controllers a priori of energy-related initiatives undertaken in their territories. An example of a community grid model is the one being developed in Austria. This process that is pursued in a country under a federalist political system can provide insights into what would appear as the final stage of community grids development after issues related to the devolution of authority in the energy sector have been addressed in countries under a centralized Jacobin political system. In these countries, the decentralization of competencies in the energy sector should necessarily involve the communalization of energy infrastructure and grid access for third parties.

Pending these reforms, planning energy systems with community grids provides local authorities with the levers required to govern their already accessible competencies that include land management and waste management. Once the legitimacy of political action is provided with these reforms, the municipalities should be provided with tools to support planning energy production systems and establishing rules that address community needs and manage collective action problems such as the reduction of greenhouse gas emissions.

In Senegal, the latest strategic framework document for the energy sector (LPDSE 2019–2023) sets the objective in terms of transition to energy sustainability at the national level (Ministere en charge des Energies, 2019). The document is unclear on the potential contributions of local communities, especially in terms of renewable energy penetration to the grid. However, communities already have tools available to contribute to the LPDSE effectively. An example is the Covenant of Mayors, which, like the Kyoto Protocol for countries, outlines the local authorities' objectives in transition to energy sustainability. As part of the Covenant, municipalities are equipped with the tools they need to carry out energy surveys and monitor activities in the field. Municipalities that are members of the Covenant are committed to tracking the amount of energy consumed in their territory along with the corresponding level of CO_2 emissions.

Beyond the political opportunity to implement structural reforms empowering local communities, a new agreement on planning energy systems with locally available resources is an opportunity for participatory democracy within communities. It provides a platform for translating policies into choices and mapping out pathways for their implementation and operationalization and, as De Jong (2011) argues, for calling the political establishment and social actors to adopt a new governance policy.

Our proposed approach to planning community-level energy systems is a dynamic, multi-step, time-defined and context-sensitive process with three key moments: preparation, implementation and monitoring.

The Preparation Phase Shall Include the Following

- Characterization of the legal and regulatory conditions of actions in the energy sector
- Identification of critical levers of action that takes into account demands and available resources
- Formulation of energy production and climate change mitigation objectives
- Definition of implementation strategies and an agenda

The preliminary phase lays out the framework for legitimate action of the local government in the transition to energy sustainability. Senegal still lacks a regulatory platform that brings together national and local governments to define energy agendas. Nonetheless, the requirement for such a platform is laid out in the national governance framework document (Ministere en charge des Finances, 2019). The document highlights the critical roles that local authorities could play in the implementation of local development plans for the energy sector and the creation of a legal framework for energy based on regulatory (conventions), strategic (local energy policy) and operational (local action plan) objectives. The preliminary phase is also an opportunity to identify key stakeholders in the community to be involved in the process of developing a shared vision of the objectives and expected results. We agree with Samb (2014) that planning a participatory local energy agenda is a protracted and critically important process that requires the identification and early involvement of key community stakeholders as advocates of local democracy.

The Implementation Phase Shall Include the Following

- Setting priorities for action
- Selection of indicators and monitoring tools among available toolkits
- Implementation of activities listed in the work plan

The second phase provides information on the effectiveness of local government action in implementing the sustainable energy agenda. During this phase, the municipality should pay closer attention to the stakeholders' engagement and limitations to define responsibilities, processes and operational methods. During this phase, actions are only legitimate if the work programme is in line with the national energy agenda. Support from international organizations such as the Covenant of Mayors can be directly channelled to local authorities through the provision of demographic, economic and energy development models. However, it is the municipality's responsibility to ensure that priorities highlighted in these documents are aligned with national priorities and that action complies with the regulatory framework in force at the national level.

The Monitoring, Verification and Evaluation Phase Shall Include Following

- Periodic evaluation of action effectiveness
- Public information and consultation with all stakeholders

The third phase provides an opportunity to monitor and report on the human, financial and technical resources deployed for the implementation of the sustainable energy agenda. This phase requires regular communications with the community on results, bottlenecks and projections. The participatory evaluation of energy infrastructure and services should support a continuous improvement of processes. As Bouvier (2003) mentions with respect to the geopolitical challenges of electricity distribution, local authorities and stakeholders (unions) need to re-establish themselves as the "custodians of the public service."

4 Conclusion

This chapter presents a new approach to defining local energy agendas, drawing on lessons learned in pioneer countries. The countries in sub-Saharan Africa have different legal and regulatory frameworks applicable to the energy sector at the local level, but actions carried out over the last decades have progressively supported a transition to energy sustainability driven by local communities. The process for empowering municipalities to take a leading role in the agenda for transition to energy sustainability is irreversible in sub-Saharan Africa as it is in other regions. Our study demonstrates the existence of legal bases for moving forward, as well as tools for supporting action. However, sub-Saharan Africa countries still lack legal frameworks that lay the foundations for a consistent agenda that reconcile local ambitions with national strategies. Additional reforms are required for municipalities to anchor these ambitions with the legitimacy of a public authority delegation and the efficiency of a public service provider for the community.

Wolfgang Streicher quoted by Sinai (2012), defines energy planning in local communities as a mechanism to engage and organize municipalities in the next years to deliver energy transition and create adequacy between the energy demands and supply with locally available resources. We have seen throughout the chapter that countries in sub-Saharan Africa need political reforms to design these islands of energy communities, which would not operate as stand-alone, but would connect to other communities to deliver the greater value of access to modern, reliable and sustainable energy (SDG-7) using local resources.

References

Bouvier, G. (2003). The geopolitical stakes of electricity distribution in France. *Herodote, 110*(3), 71–87. https://doi.org/10.3917/her.110.0071

Darghouth, N. R., Wiser, R. H., Barbose, G. L., & Mills, A. D. (2015). *Net metering and market feedback loops: Exploring the impact of retail rate design on distributed PV deployment*. Ernest Orlando Lawrence Berkeley National Laboratory. Retrieved from https://eta-publications.lbl.gov/sites/default/files/lbnl-183185_0.pdf

European Union. (2009). *EuropeanUnion commission electricity directive (2009/72/EC)*. Brussels, . Retrieved from https://eur-lex.europa.eu/eli/dir/2009/72/oj

Godinot, S. (2011, Mars). Les plans climat énergie territoriaux : voies d'appropriation du facteur 4 par les collectivités et les acteurs locaux? *Développement Durable & Territoires: Économie,*

Géographie, Politique, Droit, Sociologie, 2(1). https://doi.org/10.4000/developpementdurable.8874.

Jon Pierre, A. S. (2014). Introduction: Toward output legitimacy in local government? *Urban Research & Practice,* 7(2), 119–122. https://doi.org/10.1080/17535069.2014.910933

Jong, S. D. (2011, May 24). Towards global energy governance: How to patch the patchwork. *International Development Policy | Revue internationale de politique de développement.* https://doi.org/10.4000/poldev.736.

Ministere en charge des Collectivites Territoriales. (2013, January 17). *Acte III Decentralisation.* Consulté le September 10, 2020, sur Decentralisation.gouv.sn: http://www.decentralisation.gouv.sn/l%E2%80%99acte-iii-de-la-decentralisation

Ministere en charge des Energies. (2019). *LPDSE 2019–2023.* Consulté le September 10, 2020, sur energie.gouv.sn: http://www.energie.gouv.sn/lettre-de-politique-de-developpement-du-secteur-de-lenergie/

Ministere en charge Des Finances. (2019). *Fiche-dopportunite-sectorielle-energie.pdf.* Retrieved May 8, 2020, from Finances.gouv.sn: http://www.finances.gouv.sn/wp-content/uploads/2019/02/fiche-dopportunite-sectorielle-energie.pdf

Samb, N. (2014). *Gouvernance territoriale et participation citoyenne au Sénégal.* niversité Paul Valéry - Montpellier III, Geographie. Montpelier: HAL Archives Ouvertes. Récupéré sur https://base.afrique-gouvernance.net/docs/2014_gouvernance_territoriale_e_participation_citoyenne_senegal.pdf

Sinai, A. (2012, June 26). L'Autriche 100% renouvelable : une utopie concrète de l'après-pétrole. *Actu Environnement.* Consulté le September 2020, sur https://www.actu-environnement.com/ae/news/autarcie-energetique-autriche-vorarlberg-16024.php4#xtor=EPR-1

Chapter 3
Can Municipalities Lead the Energy Transition? On Available Policy, Competences and Requirements for Action

Cheikh Faye and P. Macharia

Abstract The implementation of innovative, bottom-up and tailor-made sustainable energy solutions in local communities is a fundamental game-changer in countries' energy transition agendas. Challenges related to energy governance and the action to mitigate global warming require these communities to play a leading role through increased involvement in local energy initiatives. Municipalities can play this role at different levels, as a bridge between the private sector, the development agencies and the local communities, throughout the sector value chain from energy production to end-use consumption, using locally available renewable energy resources. The study was carried out in the city of Dakar, Senegal, where data on energy behaviour were collected through a survey. The main conclusion drawn from the study is municipalities can lead the transition to energy sustainability with the support of resident communities. However, this requires a policy framework that defines the municipalities' scope of action, which is currently missing.

Keywords Energy transition · Policy · Municipalities · Local agenda

1 Introduction

According to the International Renewable Energy Agency (IRENA), cities consume about 65% of the global energy supply and account for about two-thirds of the global anthropogenic carbon emissions (IRENA, 2016). Therefore, besides increasing global attention focused on achieving the UN Sustainable Development Goal number 7, and the implementation of the Paris Agreement, the role of cities and municipalities in tackling climate change, promoting sustainable energy transitions

C. Faye (✉)
Ministry of Labour, Business Integration Programme, Dakar, Senegal
e-mail: cheikhfaye990@gmail.com

P. Macharia
Vienna University of Technology (TU Wien), Vienna, Austria

A. Fall, R. Haas (eds.), *Sustainable Energy Access for Communities*,
https://doi.org/10.1007/978-3-030-68410-5_3

and expanding the use of renewable energy is gaining recognition (Qudrat-Ullah et al., 2020), and (Hoppe et al., 2015). The reason is the municipalities' position as the interface between the national government, the local communities, the private sector and the international community. Empowering municipalities can accelerate the adoption of smart technologies with increased efficiency and promote energy autarky. This role can be assumed in local energy initiatives (LEIs) that support value addition and markets for locally available energy resources, bottom-up business models for job creation, and national grid load alleviation schemes (Schmidt et al., 2012).

Traditionally, municipalities are tasked with the provision of high energy-consuming services, including transport, water and wastewater management services, and buildings maintenance through closely linked systems and networks of actors providing and consuming various services at various scales. Thus, the increased interest of stakeholders, including development agencies, on the way to leverage the unique role of municipalities as large consumers of energy and as intermediaries to implement local communities' energy agendas. However, as Gustafsson and Mignon (2019) argue, this role is also quite complex, depending on the structure of national governance, whether centralized or decentralized. The authors observe that the bottom-up approach to energy transition used in decentralized governance is highly participatory, innovative and competitive. Engelken et al. (2016) demonstrate that when the municipalities leadership aims for energy self-sufficiency, this can be a catalyst for the energy transition. In addition, da Silva and Horlings (2020) observed that local energy production initiatives are important contributions to access renewable energy, especially when municipalities are involved through the provision of resources to support the promoters.

Since 2008, more than 50% of the world's population live in cities. This proportion is expected to increase to 67% by 2050 (Rosenzweig et al., 2010). This rapid urbanization will mostly take place in Africa, where eight out of ten countries with the highest rates of urbanization in the world are located (Santos et al., 2017). Therefore, municipalities in the region have an instrumental role to play in the global transition to sustainability by using locally available renewable energy resources to supply the increasing energy demand driven by urbanization. Since cities worldwide have different sizes and energy potentials, there is a wide range of initiatives to reduce greenhouse gas emissions in the energy sector by leveraging locally available resources. There are also social reasons for the cities' leading role of action for sustainability, including the fact that cities are a hotbed of change actions (Bakker, 2003). The recent citizens' mobilization has repositioned issues such as supply with renewable energy resources and efficiency of energy demand higher in the agenda of decision-makers. In sub-Saharan Africa, these social incentive and infrastructure investments should primarily target diverse challenges that still impede the implementation of local energy initiatives, including:

- Organization of a balanced supply and demand market with reliable and tailor-made technologies

- Access to financing mechanisms that facilitate access to renewable and efficient energy technologies
- Establishment of a regulatory framework that provides legitimacy to a local energy agenda
- Availability of managerial resources to implement and monitor initiatives in the field

Municipalities' participation in the implementation of local energy initiatives is crucial not only in mitigating climate change, but also in promoting local energy markets, with integrated value chains. In addition, the accelerated uptake of decentralized renewable energy systems, which is driven by increased innovation in off-grid and mini-grid technologies, contributes to reducing production costs (Ministere en charge des Energies, 2017). This provides an important opportunity for the promotion of entrepreneurship using local energy resources for clean cooking and lighting. The current situation of energy access and energy efficiency in West Africa is somewhat paradoxical. Firstly, the majority of the population has no access to modern energy services. Secondly, a significant share of the existing energy resources is wasted (ECREEE, 2013). In addition, Qudrat-Ullah et al. (2020) note that in the African context, although there are accelerated efforts towards energy access from off-grid renewable energy sources, the role of municipalities, especially in sustainable energy planning and the transition is still unsatisfactory. However, the literature evidenced the role of municipalities in the design and implementation of energy initiatives that meet the local demand with locally available resources (Hoppe & Miedema, 2020), and (Warbroek & Hoppe, 2017).

This chapter aims to introduce levers of action for municipalities in sub-Saharan Africa to drive the political agenda of the energy transition to sustainability, learning from past initiatives and current challenges. In this regard, the study aims to answer the following questions: How do consumers perceive the use of locally available energy solutions? How could municipalities lead the transition to sustainable energy production and consumption locally?

2 Methodological Approach

The study approach consisted of a structured survey conducted in Diamniadio, a district located in Dakar, Senegal, in order to understand drivers of energy behaviour in the city.

2.1 Survey Site

Diamniadio is a district in the municipality of Rufisque, which is located 30 km from the Dakar Metropolitan area (Journal Officiel Gouvernement du Senegal, 2002).

According to data from the National Statistics and Demography Agency (ANSD), the district is home to 2274 dwellings and 1205 commercial buildings in 2018. ANSD also provided data on 136 small businesses using makeshift facilities such as fruit stalls. The majority (52.6%) of surveyed dwellings have revenues between EUR 152 (CFA F 100,000) and EUR 381 (CFA F 250,000) per month. Two structured survey questionnaires, one for residential buildings and another for commercial buildings, were prepared to collect data. The questionnaire applied to residential buildings includes nine sections with questions on the characteristics of the buildings (size, envelope and orientation), various energy services (lighting, indoor cooling, cooking, etc.), energy supply options, and the perception of social and environmental parameters related to the energy sector (CO_2 emissions, odour, smoke, etc.). The questionnaire applied to commercial buildings includes questions on building characteristics, energy supply options and perception of social and environmental parameters related to the sector but excludes cooking energy from energy services.

2.2 *Sampling, Data Collection and Processing*

The survey on citizens' energy behaviour in Diamniadio was conducted between November and December 2018. The survey sample was statistically computed from ANSD data on residential and commercial buildings, considering a significance rate of 95%. The sample size was 330 residential buildings and 350 commercial buildings. The building selection used the random-step approach, which means that for each of the five data collector groups, the first building is randomly selected and the following ones are equally distanced (the step). The survey was carried out within the framework of the Sustainable Energy Access for Sustainable Cities (SEA4cities) project. The SEA4cities team, accompanied by 14 residents of the district, went door-to-door with the questionnaires in order to capture how citizens consume energy. The study seeks to understand how this behaviour affects four levers of action that municipalities can pull for transition to energy sustainability:

- Promotion of efficient energy services through the control of appliances and devices available in the local market.
- Decentralized energy production with locally available renewable energy resources through community grids or private wire networks.
- Management of sustainable energy production and efficient energy consumption through the recruitment of trained human resources who are able to plan the demand-supply equilibrium.
- Levers of communication on and dissemination of good practices to motivate citizens' contributions to the local action agenda.

The strategies proposed in this chapter derive from the results of data assessment in the background of these four pillars. The two first sections of the questionnaire target the building characteristics and the profile of occupants. Sections 3–6 of the questionnaire target the dwellings' energy consumption. Questions include energy-consuming appliances and devices in the building such as lamps, ovens and air conditioners.

In sections 7 and 8, data on the perception of building occupants of the social and environmental parameters related to energy use, such as the cost of electricity from the grid and material pollution, were gathered. The sections also include questions on the current state of access to energy with decentralized systems.

The open questions in Sect. 9 relate to the economic and environmental motivations of a hypothetical transition to a renewable energy supply such as rooftop solar photovoltaic installations or waste recycling units. There, we question the interviewees' awareness of global challenges related to the energy sector, such as the mitigation of climate variability, reduction of reliance on oil imports, and impact of greenhouse gas emissions on sea-level rise that affects many districts in Dakar, including the adjacent districts of Bargny and Malika. The Excel-based analysis of data collected from the survey was completed by a review of literature on energy in urban environments.

3 Results and Discussions

The survey response rate was 115% in residences (368 surveyed) and 63.4% in commercial buildings (222 surveyed).

3.1 Energy Services

The Excel-based analyses of data collected were grouped in three (3) indicators:

- Energy behaviour with respect to major services including lighting, ventilation and other plugging appliances in buildings
- Energy supply with respect to connection to the grid or alternative supply options using diesel or renewable energy resources in buildings
- Demand-supply equilibrium at the level of the community

Ventilation in buildings accounted on average for 51.42% of electricity consumption in the buildings surveyed, while lighting accounted for 19%. The rest of the electricity is consumed by other plugging appliances and devices, which include fridges, televisions and mobile phones. Therefore, efforts to improve energy use in buildings should dedicate a higher share of attention to ventilation devices and modalities to reduce their energy intensity. Action in this domain can include legal provisions in the building code regarding materials and architecture and restrictions in the market that set efficiency threshold (in BTU) for devices such as (air conditioners and fans), considering also cost for communities. The regulator can set the market signal with a recommendation list of devices having a weighted score combining efficiency and market price.

3.2 Supply Options

All surveyed households were connected to the national electricity grid, and one had an additional solar photovoltaic system. The analysis of waste-to-energy as an alternative to the grid is because there are previous studies on its potential in the district, which confirm readily available resources. Data in the analysis are from the report of *Unité de Coordination et de Gestion des déchets* (UCG), the municipality department in charge of waste management (Unite de Coordination et de Gestion des Dechets, 2016). Between 2014 and 2015, UCG completed a waste characterization study in the district of Diamniadio. Three (3) indicators are considered in the assessment of the potential for recycling buildings waste into energy:

- Quantity of waste generated in buildings (in kg/year)
- Energy potential of buildings waste converted through gasification (in kWh/year)
- Energy potential of buildings waste converted through incineration (in kWh/year)

Figure 3.1 shows the potential of energy recovery from the waste produced in buildings in Diamniadio using two waste-to-energy technologies: incineration and gasification. The potential for recovering energy from waste through gasification is established based on the proportion of biodegradable materials in waste collected (13%). Thus, the average potential is projected at 1.664 kWh per year. The recovery potential by incineration is established based on the proportion of waste materials that can be incinerated, including wood logs, plastics and paper waste (46.7%). The average potential is 5978 kWh per year.

Thus, it appears that energy recovery by waste incineration returns a higher energy potential. This information on the potential of energy recovery from available waste is necessary for planning periods where energy recovered from waste in community grids or stand-alone systems is enough to meet the demand and periods

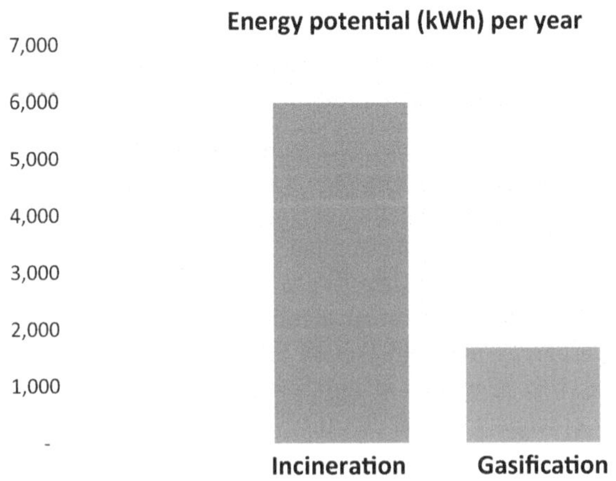

Fig. 3.1 Potential of energy recovery from waste generated by surveyed buildings

where supply from the interconnected grid is necessary. Since the intermittence factor is less important in waste to energy than it is in solar photovoltaic and wind energy, such planning with energy potential recovered per period could provide better predictions on the quantity of energy required from the grid.

3.3 Energy Supply and Demand in Local Communities

In Senegal, the main source of energy supply in cities is the interconnected grid, with over 90% of city buildings connected to a grid (International Energy Agency, 2020). The interconnected grid is characterized by the prevalence of fossil fuels in electricity production. Other energy uses in buildings, such as liquefied petroleum gas in cooking, also rely on fossil fuels, namely crude oil. In 2016, diesel oil, heavy fuel and natural gas accounted for over 80% of the overall country's energy supply. Senegal averaged about 48% energy independence (Ministere en charge des Energies, 2018). The rate has improved slightly since, due to the progressive introduction of large solar power plants in the grid generation.

However, the situation also depicts the underutilized potential of locally available energy resources to reduce both dependence on imported fossil fuels and greenhouse gas emissions as a contribution of local communities to the global action for mitigating climate change. This situation is unfortunate because it fails to integrate a powerful driver of communities' ownership of the energy agenda. Two points are important to consider here.

Firstly, the survey in Diamniadio shows that the social value of energy matters to citizens. Figures 3.2, 3.3 and 3.4 lists the various attributes associated with energy fuels by surveyed populations and their degree of significance.

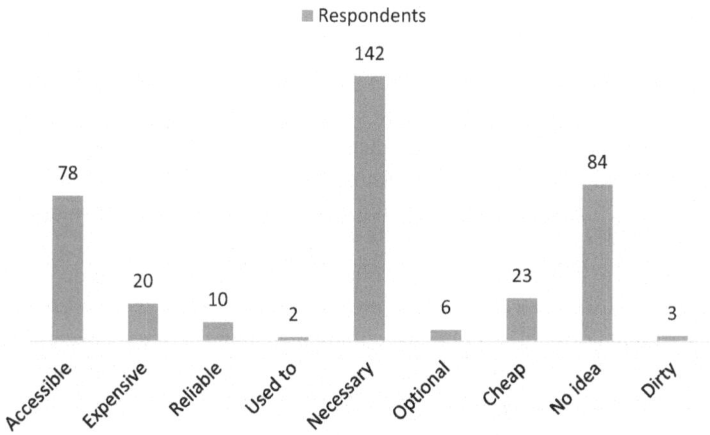

Fig. 3.2 Attributes associated with Liquefied Petroleum Gas (LPG)

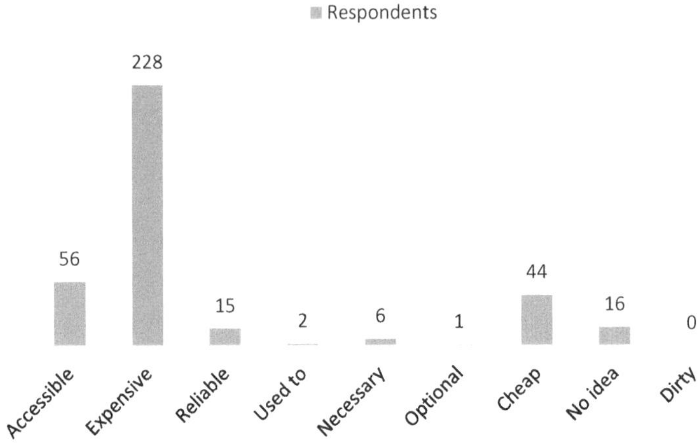

Fig. 3.3 Attributes associated with electricity

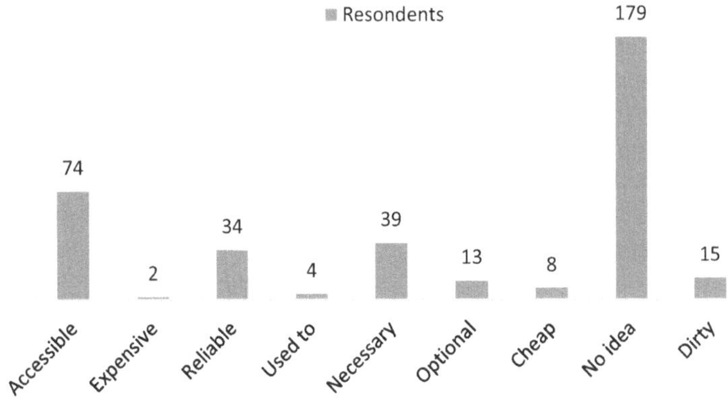

Fig. 3.4 Attributes associated with Charcoal

The majority of respondents (38.6%) associate liquefied petroleum gas (LPG) with the attribute "Necessary," followed by 22.8% of respondents who do not express an opinion on this fuel (No idea). The proportion of respondents who associate LPG with the attribute "Accessible" is also relatively important (21.2%).

The majority of respondents (62%) associate electricity with the attribute "Expensive," and only 4% of respondents do not express an opinion about this form of energy (secondary energy produced from primary fuels and intended to power another use). This suggests a specific interest in electricity that was observed during the survey. The majority of respondents complained about their expensive electricity bills. Still, 12% of respondents associate electricity with "Cheap." This may relate to a problem of communication between the interviewer and the respondents.

The majority of the respondents (48.6%) did not express an opinion on charcoal, and 21% of the respondents associate the fuel with "Accessible."

Secondly, the respondents are aware of the environmental impact of pollution from energy use, because effects related to global warming are visible at the community level. For instance, sea-level rise primarily affects fishermen in the coastal area of Dakar who lost their houses, and therefore, their jobs in many cases. This absence of revenues affects municipality taxes and threatens the existence of the municipality when people are forced to settle outside the traditional community.

3.4 Determinants of Energy Efficiency in Local Communities

Energy efficiency is another important pillar of the energy transition agenda where municipalities have levers of action. The survey data show that actions to improve the efficiency of energy use in local communities are impeded by four factors.

Firstly, the financial capacity of households and commercial buildings is not enough to invest in retrofit initiatives with efficient appliances and devices. Some segments of the population surveyed, mainly in the residential sector, continue to focus more on economic parameters on the selection of plugging devices. The residential sector predominantly uses less efficient, second-hand or low-technology appliances that feature poor energy performances.

Secondly, coordination between local and national stakeholders (political decision-makers, industrialists, local residents) on energy planning is limited. Renewable energy penetration targets and energy-saving objectives are usually decided at the national level, and when it comes to implementation at the local level, there is either a lack of ownership or a deficit of resources to implement them. More than often, neither ownership nor resources are available to implement locally. And the political framework of possible actions remains loose. All competencies transferred to local authorities by the 1996 Decentralization Act include an energy dimension that is not fully addressed in action agendas. One example is the transfer of waste management competencies to local governments whose framework of application overlooks the potential of recycling this waste to energy, as demonstrated by the results of our survey. This situation is compounded by the absence in small municipalities such as Diamniadio of qualified resources that can reflect this energy dimension in documents from the central governments and initiate a bottom-up approach, rather than just a top-down approach.

The lack of financial and human resources in municipalities impedes the development of incentive mechanisms. Investments on renewable energy technologies and energy-efficient appliances and devices feature the specificity of important resources required upfront and less during operation. Then, it is important to design mechanisms that take into account the need to mobilize human and financial resources at early stages. Such mechanisms require the involvement of capital assets managers such as banks or community financing mechanisms. An example in

Senegal is tontines that are managed by women's associations. Mechanisms that use this readily available capital to fund the energy transition are possible, and each municipality can be innovative, depending on its social context and energy market.

3.5 Pillars of a Local Transition to Energy Sustainability

Learning from the results of our survey, four pillars necessary in the implementation of a sustainable energy agenda at the local level were identified:

Definition of a clear action agenda that complies with national objectives. National legislation on renewable energy and energy efficiency, including objectives, should serve as a background to design a local energy strategy that includes the following:

- Identification of the energy behaviour of the community with data on energy-consuming devices and appliances, modalities of consumption such as comfort temperature and lighting requirements in lumen per square metre.
- Identification of supply options, including locally available resources such as waste materials that could be converted to energy.
- Definition of objectives and achievable indicators related to these objectives on the transition to energy sustainability at the local level. Thus, it is important to identify priority areas of action such as ventilation in the surveyed district, to achieve higher eco-efficiency factors for the municipality action.
- Identification of bottlenecks and areas where the municipality's involvement is required to push the transition agenda, including the involvement of capital assets managers, labelling devices available in the market to inform consumers on quality, and contribution to infrastructure projects and programmes (e.g. accessibility of lands) as an investment incentive that could contribute to lower financial risks.

Development of appropriate tools to track and monitor progress towards the objectives set out in the energy transition agenda.

Communication on the strategy: the municipality has the capacity to mobilize all players at the local level to support its action. This mobilization can include information meetings, training youths of the community in manufacturing the spare parts of the technology that can be manufactured locally and training on the maintenance of installed technologies.

Management of day-to-day activities of the agenda implementation: the management of an energy transition agenda at the local level should include, without limitations, the activities listed below:

- Periodically collect energy data in community buildings in order to continuously refine planning models and update the objectives.
- Initiate and implement demonstration activities in community/public buildings, such as lighting retrofit and passive ventilation. The municipality could also lead

by example through the adoption of low-carbon measures in the transport of its personnel (use of mass transport, car-sharing, bicycle, etc.).
- Integrate energy efficiency and ecological criteria in public procurement. The municipality can include energy indicators in its criteria for the selection of public service suppliers, for instance, energy performance of office equipment, in addition to existent criteria required by the Code des Marches Publics.
- Recruit qualified human resources for the inspection of new buildings' compliance with the energy efficiency requirements in the 2009 Building Code.

4 Conclusion

Over the years, Senegal has developed a comprehensive policy towards sustainability. The country has the potential to achieve universal access to energy services, energy self-sufficiency—excluding biomass—and a diversified energy mix by 2030. However, the question of what could the role of municipalities be in this context remains unanswered, especially at a time when the decentralization of national competencies is an important component of the central government policy. In 2013, the country started the third phase (Act 3) of its decentralization process. The process, so far, overlooked the potential contribution of municipalities in the energy sector that is cross-cutting to all sectors of decentralized competencies. The chapter demonstrated that municipalities could have a significant role in design and implementation of the national energy agenda, which could reconcile the objectives of the country in terms of renewable energy penetration, climate change mitigation and political decentralization.

References

Bakker, K. J. (2003). A political ecology of water privatization. *Studies in Political Economy*, 35–58. https://doi.org/10.1080/07078552.2003.11827129

ECREEE. (2013). *Politique d'Efficacite Energetique de la CEDEAO*. Praia, Cape Verde. Consulté le June 1, 2020, sur http://www.ecreee.org/sites/default/files/documents/basic_page/politique_defficacite_energetique_de_la_cedeao.pdf

Engelken, M., Römer, B., Drescher, M., & Welpe, I. (2016, November). Transforming the energy system: Why municipalities strive for energy self-sufficiency. *Energy Policy, 98*, 365–377. https://doi.org/10.1016/j.enpol.2016.07.049

Gustafsson, S., & Mignon, I. (2019). Municipalities as intermediaries for the design and local implementation of climate visions. *European Planning Studies, 28*(6), 1161–1182. https://doi.org/10.1080/09654313.2019.1612327

Hoppe, T., Graf, A., Warbroek, B., Lammers, I., & Lepping, I. (2015, February). Local governments supporting local energy initiatives: Lessons from the best practices of Saerbeck (Germany) and Lochem (the Netherlands). *Sustainability, 7*(2), 1900–1931. https://doi.org/10.3390/su7021900

Hoppe, T., & Miedema, M. (2020). A governance approach to regional energy transition: Meaning, conceptualization and practice. *Sustainability, 12*(3). https://doi.org/10.3390/su12030915
International Energy Agency. (2020). *World energy outlook 2020.* IEA. Retrieved from https://www.iea.org/reports/world-energy-outlook-2020
IRENA. (2016). *Renewable energy in cities.* International Renewable Energy Agency. Retrieved from https://www.irena.org/publications/2016/Oct/Renewable-Energy-in-Cities
Journal Officiel Gouvernement du Senegal. (2002, February 21). *DECRET n° 2002-171 du 21 févier 2002 portant création des communes de Diamniadio, Cayar, Mboro, Ranérou, Rosso-Sénégal, Niandane et Diawara.* Dakar, Senegal. Consulté le June 2, 2020, sur https://web.archive.org/web/20180326064402/http:/www.jo.gouv.sn/spip.php?article1872
Ministere en charge des Energies. (2017, July 19). *Expérience du Sénégal sur les Concessions d'Electrification Rurale.* Dakar, Senegal. Consulté le June 2, 2020, sur http://www.ecreee.org/sites/default/files/editor/expeience_senegal_sur_les_concessions.pptx.pdf
Ministere en charge des Energies. (2018). *Rapport SIE 2016.* Ministere en charge des Energies.
Qudrat-Ullah, H., Akrofi, M. M., & Kayal, A. (2020, April). Analyzing actors' engagement in sustainable energy planning at the local level in Ghana: An empirical study. *Energies, 13*(8). https://doi.org/10.3390/en13082028
Rosenzweig, C., Solecki, W., Hammer, S. A., & Mehrotra, S. (2010). Cities lead the way in climate–change action. *Nature, 467*, 909–911. https://doi.org/10.1038/467909a
Santos, S. D., Adams, E. A., Neville, G., Wada, Y., Sherbinin, A. D., Bernhardt, L. M., & Adamo, S. B. (2017). Urban growth and water access in sub-Saharan Africa: Progress, challenges, and emerging research directions. *Science of the Total Environment, 607-608*, 497–508. https://doi.org/10.1016/j.scitotenv.2017.06.157
Schmidt, J., Schönhart, M., Biberacher, M., Guggenberger, T., Hausl, S., Kalt, G., … Schmid, E. (2012, August). Regional energy autarky: Potentials, costs and consequences for an Austrian region. *Energy Policy, 47*, 211–221. https://doi.org/10.1016/j.enpol.2012.04.059
Silva, D. S., & Horlings, L. G. (2020). The role of local energy initiatives in co-producing sustainable places. *Sustainability Science, 15*, 363–377. https://doi.org/10.1007/s11625-019-00762-0
Unite de Coordination et de Gestion des Dechets. (2016). *Campagne de Caracterisation des Dechets 2014-2015: Diamniadio.* UCG. Dakar.
Warbroek, B., & Hoppe, T. (2017). Modes of governing and policy of local and regional governments supporting local low-carbon energy initiatives; exploring the cases of the Dutch regions of Overijssel and Fryslân. *Sustainability, 9*(1). https://doi.org/10.3390/su9010075

Chapter 4
Local Action for Energy Sustainability: A Review of Policies' Impact

C. A. Mbodji, C. Sumereder, D. Diouf, and A. Maiga

Abstract Energy sustainability in practice demonstrates the need to scale down the focus from the national to the local level. The literature proposes different models and policies for implementing sustainable energy solutions in communities. In this chapter, the authors review the policy competencies of local governments and propose a new model for transition to energy sustainability in communities, learning from different initiatives. The study confronts this model with lessons learned in transitioning to energy sustainability in four sub-Saharan African countries: Ghana, Nigeria, Mozambique and Senegal. The main findings are countries made significant efforts to improve energy access in rural communities with renewable energy systems and to improve the efficiency of energy use in urban communities. However, prerequisites such as investment in the network infrastructure and in planning tools, especially at the scale of municipalities, continue to impede progress in access and transition to energy sustainability.

Keywords Sustainable energy · Energy policy · Local action

1 Introduction

Energy plays an instrumental role in economic development. It supports the emergence of developing countries and the eradication of poverty through the development of new economic value chains. Fossil fuels still account for the largest share of global primary energy supply, with 81% in 2018, including 32% of fuel oil

C. A. Mbodji (✉) · D. Diouf · A. Maiga
Laboratoire Electronique, Informatique, Télécommunications et Energies Renouvelables, Gaston Berger University, Saint-Louis, Senegal
e-mail: mbodji.cheikh-ahmadou@ugb.edu.sn; djicknoum.diouf@ugb.edu.sn; amadou-seidou.maiga@ugb.edu.sn

C. Sumereder
FH Joanneum, University of Applied Sciences, Kapfenberg, Austria
e-mail: christof.sumereder@fh-joanneum.at

A. Fall, R. Haas (eds.), *Sustainable Energy Access for Communities*,
https://doi.org/10.1007/978-3-030-68410-5_4

(International Energy Agency, 2020a, b). Different scenarios of the global demand for primary energy predict an increase between 27% and 61% by 2050 (World Energy Council, 2013). Transition to energy sustainability in this context requires not only technology initiatives in both supply and demand within the sector but also in policymaking. Energy policies are of paramount importance in driving sustainability action. At the international level, global commitments to advance the energy sustainability agenda, namely the Sustainable Development Goals (SDGs) and the Nationally Determined Contributions of countries (NDC), following the climate agreements of COP21 and COP22, are meant to shape the process of this transition. At the national level, innovation in energy legislation should anchor achievements of the last years in terms of renewables penetration into the grid and measures to improve the efficiency of energy use, especially in the sub-Saharan Africa region (Mandelli et al., 2014). These legislations should also open a new era of transition to energy sustainability driven by local communities. Several authors, including Davidson et al., 2007, call for a transition to sustainability using locally available renewable energy resources combined with energy efficiency measures. Energy efficiency is defined as essentially using less energy to provide the same service (Rosen, 2009). However, challenges exist in the way to this transition.

The energy challenges in sub-Saharan Africa are numerous and affect the overall performance of the region's social and economic indicators. Ganda and Ngwakwe (2014) argue that major problems in promoting sustainable energy in sub-Saharan Africa are: (1) continued fossil fuel subsidies; (2) presence of monopoly structures in the energy sector; (3) lack of seed capital; (4) high transaction costs; and (5) risk on low-carbon finance defined as the vulnerability of investments intended to sustainable energy infrastructure to become reversed or modified. Some of these problems could be alleviated by scaling down the scope of interventions to address the problems in a local context, in particular those related to capital and technology investments in the region.

In this chapter, we analyze different policies and strategies available in the literature as models used to achieve energy sustainability in local communities. From this analysis, we derive a model for local transition to energy sustainability in developing countries within the sub-Saharan Africa region. Our model is a compilation of strategies and pathways that local governments could consider in order to achieve energy sustainability, especially in rural communities. The study's objective is to advance the reflection on energy sustainability from the status of knowledge and shed light on the perspectives of local action for energy sustainability in sub-Saharan Africa.

2 Methodological Approach

Sustainable energy development requires access to energy services, which are affordable and preserve the environment. Our methodology is the literature review, from which we derive new proposals that should support policymakers in the design of strategies that promote access to energy sustainability in local communities of the sub-Saharan Africa region.

Oyedepo (2012) argues that renewable energy and energy efficiency are two components that should go together to achieve sustainable development in Nigeria. His approach involves the exploitation of locally available renewable energy resources combined with the implementation of energy efficiency measures targeting in priority the sectors of construction, industry, residential and office buildings, and transportation.

The World Energy Council (2020) analyses energy sustainability through 23 indicators, which are related to the following three dimensions relevant at both the national and local levels:

- Energy security that includes effective management of primary energy supply, reliability of energy infrastructure and ability of energy providers to supply the demand.
- Equity in the accessibility and affordability of energy supplies across the populations.
- Environmental sustainability that proceeds from improved efficiency in the supply and demand sides of the energy sector and from the development of energy infrastructure using renewable and other low-carbon sources.

The International Energy Agency (2019) proposes the Sustainable Development Scenario (SDS), which outlines an integrated approach to question the issues of energy and sustainable development. For universal access to modern energy, the scenario proposes a focus on electrification schemes, both on-grid and off-grid, and the promotion of clean cooking facilities in bringing the number of people without access to modern energy down to zero by 2030.

Mainali et al. (2014) propose the Energy Sustainability Index (ESI), which is a tool that helps policymakers and energy planners to evaluate the status and progress of sustainable energy initiatives in rural settlements in developing countries. The index features 13 indicators (techno-economic, environmental and social) that are appropriately designed to capture the dimensions of energy sustainability in rural communities.

3 The New Approach for Local Transition to Energy Sustainability

The methods mentioned above share the core objective of a transition that progressively substitutes fossil fuels with renewable energy resources in the supply side of the energy sector. Therefore, our new approach proposes to start with policies that support investment in locally available renewable energy resources by taxing conventional energy production at rates similar to that applied to other goods. Indeed, by phasing out subsidies on conventional energy production, the amount of subsidies needed to make decentralized renewable energy generation attractive is reduced. In a similar approach, Ouyang and Lin (2014) demonstrated that the negative externalities of China economic growth could be reduced from 4.46% to 0.43% if 10% of fossil fuel subsidies were removed. Another model of tax incentives is waiving taxes on technology imports. In Senegal, this second model of tax incentives has proven efficiency on investments in renewable energies. Indeed, the share of solar photovoltaic in Senegal increased from 0.01% in 2007 to 0.7% of the total installed power in 2012 (Ministere en charge Energies, 2015). In 2019, Senegal fed into the electricity grid solar power plants with a total capacity of 102 MW; another 40 MW solar park and a 150-MW wind power plant are under construction. Still, Senegal and other sub-Saharan Africa countries could go further in tax incentives with a third option similar to the Polish approach in the coal sector. Since 2012, the Polish government introduced an excise on coal where revenues are directed to fund the transition away from coal (Gençsü & Zerzawy, 2017).

Secondly, alongside initiatives that use locally available renewable energy resources as substitutes to fossil fuels, policies should target energy efficiency measures in specific sectors that include transport, industry and the residential sector at the local scale, namely at the scale of cities and rural communities. These measures can range from direct contributions to retrofit electric appliances and devices with more efficient products such as LED in lighting to ban legislations (e. g municipal orders) on inefficient products. The appropriate measures are dependent on the level of development and should primarily target non-vital services. The definition of vital services may vary across regions. However, any service that is not related to the food supply, health or public safety may be considered in the category of non-vital services. The efforts of municipalities can build from the experience of national agencies. In Senegal, the National Agency for Energy Saving and Management efficiently contributed to the ban on incandescent lamps, which targeted savings of a 70 MW power during its pilot phase (Journal Officiel Gouvernenment Senegal, 2011). This agency also implements the Monitoring, Analysis and Reduction of Electricity Expenditures in the Public Sector (SARDEL) programme, which supports local communities in improving the energy performance of their facilities and equipment. The implementation of the recommendations resulting from the energy diagnosis contributed to savings of 31.01 GWh between 2015 and 2018.

Thirdly, the effective promotion of renewable energy policies and efficiency measures at the local level should be supported by an adaptive tool made up of indicators, including social, economic and environmental indicators, which supports local authorities in monitoring their energy agenda. These indicators should include:

- The share of locally available renewable resources in the energy supply mix; this indicator should be associated with the index of dependency on energy import.
- The electricity use per capita; average electricity consumption, excluding high income, in sub-Saharan Africa was about 481.2 kWh per capita in 2014, which is relatively low compared to the World average electricity consumption per capita estimated at 3131.3 kWh (World Bank, 2020). This indicator should be

associated with the quantities of fuels in the grid supply mix, including renewable energy resources, and the quantity of electricity generated per period.

- The share of the rural population with electricity access which is already tracked in the majority of sub-Saharan Africa countries and is commonly named rural electrification rate. This indicator should be associated with social and economic criteria such as the regional gross product, the energy intensity and the index of attractiveness to national and foreign investments.
- GHG emissions from the energy sector; this indicator should be associated with an inventory of pollutants specific to the energy mix of the community (e.g CO_2, CH4, NOx).
- The share of the rural population with access to clean cooking energy; this indicator should be associated with the surface of green areas, the rate of deforestation and the demands of biomass and LPG fuels.
- The local forest coverage; this indicator should be associated with the annual rate of change in forests and other green areas and to the potential of carbon sink. The monitoring of this indicator should support an afforestation policy that is complementary to the energy agenda.

This adaptive tool, made up of social, economic and environmental indicators, can be designed in a software format, the content of which would be tailored to the needs of each community. This tool should provide functionalities to monitor initiatives for energy sustainability as inputs to the software and the resulting impact on indicators as outputs.

4 Discussion of Findings

In this section, we discuss how the above three-pronged method could complement actions in the sector that promote energy sustainability at the community level, taking the examples of Ghana, Nigeria, Mozambique and Senegal.

The Republic of Ghana adopted a Sustainable Development Plan in 2017, which integrates the SDGs with priorities on economic, social and environmental sectors. The Plan integrates, among others, the objective to develop a viable and competitive private sector and to invest in innovative and modern technologies. This objective conceptualized in the energy sector complies with our recommendation to develop policies that promote the development of energy infrastructure, using locally available renewable energy resources, which could be sustained by investment in key segments of the energy market value chain identified with information technologies. The environment component of the Plan also includes the objective to reduce the negative externalities of industrial and small-scale mining activities and to expand the use of clean energy (Government of Ghana, 2019). The adaptation of environmental criteria in the third model axis to account for emissions from the various mines can contribute to creating an integrated monitoring system in communities that host these industries in Ghana.

Similar to Ghana, the Republic of Nigeria mainstreamed SDGs in its National Development Planning (NDP) that includes the Economic Recovery & Growth Plan (ERGP 2017–2020). The ERGP focuses on economic, social and environmental dimensions of local development (Government Federal Republic of Nigeria, 2020). In addition, Nigeria undertook to realign its National Statistical System (NSS) with the requirements and indicators of the SDGs, which provides an opportunity to define indicators for the energy sector that integrate all dimensions of sustainability. At the local level, the identification of relevant energy indicators is adequately supported by the adaptive tool that makes it possible to monitor social, economic and environmental indicators related to SDGs 7, 11, 12 and 13, and therefore to monitor a sustainable local energy transition.

Mozambique is a net exporter of electricity, with more than 73% of the electricity produced being exported to South Africa (UNEP RISØ, 2013). However, about 67% of the population live in rural areas, and 78% of this population do not have access to electricity (International Energy Agency, 2020a, b). The situation is slightly better in urban areas where 52 of the population have access to electricity from the grid. This paradox is explained by the absence of network infrastructure to connect settlements located far from the grid. The use of locally available renewable energy resources such as hydropower is a competitive alternative to the central grid. Mozambique has a large hydropower potential, estimated at 14,000 MW. Hydropower accounts for more than 99% of the power produced in the country. Over the last decades, the government made sustained efforts to promote the use of other renewable energy resources. The reform of the energy sector opened the electricity sector to new entrants (1997). The reform also established the Conselho Nacional de Electricidade (CNELEC) as the regulator of the sector, and the rural electrification fund (FUNAE). Measures designed to improve the efficiency of energy use should complement these efforts to exploit locally available renewable energy resources. Monitoring the effectiveness of the renewable energy and energy efficiency initiatives with the tool indicators of an adaptive could make the energy transition accessible to remote rural communities.

Senegal adopted, in 2013, the legal framework of action in local administrations called Act 3 of Decentralization. Its aim is to organize the sustainable development agenda by 2022 in local territories (Ministere en charge des Collectivites Territoriales, 2013). The Act focuses on two strategic directions that include improving institutional and local governance and supporting partnership and financing mechanisms in local development actions, including in the energy sector. The first and third axes of our model are in line with these two orientations. Act 3 promotes local governance, and the model proposes policies to organize demand and supply in the sector that target local users. Further, our model proposes that each community be given the possibility to define its energy agenda based on the potential of energy resources available, which complies with the objective of participative democracy that the Act ultimately targets. The monitoring tool proposed for the energy sector can easily be integrated into the scope of local development statistics provided by the Decentralization Act.

From the previous examples, it appears that developing countries in sub-Saharan can have innovative pathways for sustainable development, building from what is currently available. The region is endowed with high potential for renewable energy resources, whether it is solar or hydropower. Innovation in the energy sector could consist in designing local energy agendas that are consistent with national ambitions and international commitments. This approach echoes the principles of participative democracy and inclusiveness. As in other sectors, the approach in the energy sector will need methodologies and tools to create consistency of action between the local and supra-local scales. The model proposed with three axes is flexible enough to create this consistent dynamic.

The energy transition driven from local communities should encourage these communities to define their energy agenda independently and to access mechanisms to fund this agenda. In Senegal, the financing of local initiatives is mainly from transfer funds that the central government grants to local communities, through the Endowment Fund for Decentralization (FDD) to support the effectiveness of policy action, and the Equipment Fund for Local Communities (FECL) to support investments in infrastructure (Ministere en charge de l'Environnement, 2015). A subsequent axis of this apparatus should focus on encouraging the private sector to develop mechanisms that leverage from this contribution to create markets for energy products and derivative products such as emissions credit trades (ECTs). The localized governance of the energy sector shaped around the three axes of our model should establish the framework to mobilize this private sector and support profitable investments in technology and institutional capacity building.

5 Conclusion

Sub-Saharan Africa region is home to approximately 1.2 billion people, only 48% of whom have access to electricity (International Energy Agency, 2020a, b). Population growth forecasts predict that the population can reach 2.2 billion people by 2040. The International Energy Agency also foresees a doubling of the primary energy demand of the region by 2040, compared to 2018 levels. These figures call for urgent actions in the energy sector. Access to energy empowers the supply in communities of vital services such as harvesting and food transformation, schooling and medical care. Action for access and transition to energy sustainability in these communities requires national and regional legislation complemented by the financial support of the private sector. Academia can also contribute to addressing the challenge by providing the necessary expertise required to identify indicators of sustainability and participative democracy in the sector.

In this chapter, some strategies developed by governments in order to address energy poverty in different countries of the region were presented. Besides the energy situation in the exemplified countries, theoretical and practical transition models were documented. The analysis revealed some common trends in energy legislation over the last 30 years:

- Conceptualization of security supply that would use locally available renewable energy resources in the grid
- Political willingness to keep centralized regulation of the sector, and
- Investments in rural electrification

The comparison of electrification rates over the years shows a slight improvement of energy access in rural areas, but it is unclear whether the centralized regulation scheme did improve either access or security of supply. Therefore, it appears necessary to review the agenda, at least as far as its mechanisms are concerned. The goals of universal access to electricity and other modern energy supplies set by the current agenda cannot be achieved without addressing the more pressing issues of investment in network infrastructure and technologies that incentivize and support local communities in efforts to transition towards more sustainable energy behaviours. Information technologies can provide this support with readily accessible tools available to communities in planning their energy systems. Participative actions such as survey or public consultations can inform behaviours and provide a solid foundation for an energy agenda that is driven by communities and monitored with adaptive information technologies.

References

Davidson, O., Chenene, M., Kituyi, E., Nkomo, J., Turner, C., & Sebitosi, B. (2007). Sustainable energy in sub-Saharan Africa. : International Council for Science. Retrieved May 2020, from https://www.compete-bioafrica.net/publications/publ/Sustainable%20Energy%20Report.pdf

Ganda, F., & Ngwakwe, C. C. (2014). Problems of sustainable energy in sub-Saharan Africa and possible solutions. *Mediterranean Journal of Social Sciences, 5*(6), 2039–2117. https://doi.org/10.5901/mjss.2014.v5n6p453

Gençsü, I., & Zerzawy, F. (2017). *Phase-out 2020: Monitoring Europe's fossil fuel subsidies*. Retrieved from https://www.odi.org/publications/10929-monitoring-europes-fossil-fuel-subsidies-poland

Government Federal Republic of Nigeria. (2020). *2nd Voluntary National Review on sustainable development*. Government Federal Republic of Nigeria. Retrieved from https://sustainabledevelopment.un.org/index.php?page=view&type=30022&nr=2401&menu=3170

Government of Ghana. (2019). *Voluntary National Review (VNR) report on the implementation of the 2030 agenda for sustainable development*. Government of Ghana. Retrieved October 12, 2020, from https://sustainabledevelopment.un.org/content/documents/23420Ghanas_VNR_report_Final.pdf

International Energy Agency. (2019). *World energy model*. IEA. Retrieved June 2020, from https://www.iea.org/reports/world-energy-model/sustainable-development-scenario

International Energy Agency. (2020a). *Key world energy statistics 2020*. IEA. Retrieved from https://www.iea.org/reports/key-world-energy-statistics-2020

International Energy Agency. (2020b). *World energy outlook 2020*. IEA. Retrieved October 7, 2020, from https://www.iea.org/reports/world-energy-outlook-2020

Journal Officiel Gouvernenment Senegal. (2011, January 28). *Décret n° 2011–160 du 28 janvier 2011*. Dakar, Senegal. Consulté le May 2020, sur http://www.jo.gouv.sn/spip.php?article8800

Mainali, B., Pachauri, S., Rao, N. D., & Silveira, S. D. (2014, April). Assessing rural energy sustainability in developing countries. *Energy for Sustainable Development, 19*, 15–28. https://doi.org/10.1016/j.esd.2014.01.008

Mandelli, S., Barbieri, J., Mattarolo, L., & Colombo, E. (2014, September). Sustainable energy in Africa: A comprehensive data and policies review. *Renewable and Sustainable Energy Reviews, 656-686*. https://doi.org/10.1016/j.rser.2014.05.069
Ministere en charge Collectivites des Territoriales. (2013, January 17). *Acte III Decentralisation*. Consulté le May 13, 2020, sur Decentralisation.gouv.sn: http://www.decentralisation.gouv.sn/l%E2%80%99acte-iii-de-la-decentralisation
Ministere en charge Energies. (2015). *Plan d'Actions National des EnergiesRenouvelables (PANER) 2015–2020/2030*. Dakar: Ministere en charge Energies. Consulté le June 2020, sur https://www.se4all-africa.org/fileadmin/uploads/se4all/Documents/Country_PANER/Senegal_Plan_d_Actions_National_des_Energies_Renouvelables_.pdf
Ministre en charge de l'Environnement. (2015). *Lettre de Politique du Secteur de l'Environnement et du Developpement Durable 2016–2020*. Dakar: Ministere en charge Environnement. Consulté le June 20, 2020, sur https://chm.cbd.int/api/v2013/documents/A7562A6D-A495-3267-4BB5-8B727A4646A3/attachments/lettre%20de%20politique%202016-2020.pdf
Ouyang, X., & Lin, B. (2014). Impacts of increasing renewable energy subsidies and phasing out fossil fuel subsidies in China. *Renewable and Sustainable Energy Reviews, 37*, 933–942. https://doi.org/10.1016/j.rser.2014.05.013
Oyedepo, S. O. (2012). Energy and sustainable development in Nigeria: The way forward. *Energy Sustainability and Society, 12*. https://doi.org/10.1186/2192-0567-2-15
Rosen, M. A. (2009). Energy sustainability: A pragmatic approach and illustrations. *Sustainability, 1*(1), 55–80. https://doi.org/10.3390/su1010055
UNEP RISØ. (2013). *Emissions Reduction Profile: Mozambique*. UNEP.
World Bank. (2020). *Electric power consumption (kWh per capita)*. US. Retrieved November 2020, from https://data.worldbank.org/indicator/EG.USE.ELEC.KH.PC?locations=ZF
World Energy Council. (2013). *World energy scenarios: Composing energy futures to 2050*. World Energy Council. Retrieved May 13, 2020, from https://www.worldenergy.org/assets/downloads/World-Energy-Scenarios_Composing-energy-futures-to-2050_Full-report1.pdf
World Energy Council. (2020). *World energy trilemma index 2020*. World Energy Council. Retrieved from https://www.worldenergy.org/publications/entry/world-energy-trilemma-index-2020

Chapter 5
Techno-Economic Assessment of Renewable Energy Potential in Cities: Case Studies of Solar Photovoltaic, Waste-to-Energy and Wind Energy

A. Fall and R. Haas

Abstract Theoretically, all places on earth are endowed with renewable energy resources. However, the requirements for co-existence with existing energy and non-energy infrastructure limit the technical realization of these resources. Scientific publications on the renewable energy potential of cities mainly rely on geographic information system (GIS) data such as solar and wind maps. However, planning systems that realize the potential of renewable energy resources need additional tools beyond GIS. In this study, we consider two additional dimensions in the assessment of cities' renewable energy potential: competing space and competing resources. The findings show the impact of city land-use density, infrastructure and capital cost on the achievement of a transition to the energy sustainability objective with locally available renewable energy resources.

Keyword Solar photovoltaic · Waste-to-energy · Wind energy · Transition costs

1 Introduction

For (Wolsink, 2018) the energy available from renewable sources is abundant, but the real scarcity derives from the space needed for the infrastructure. Natural landscape and principles of co-existence with existent energy and non-energy infrastructure limit these resources' technical realization. The scientific publications on cities' renewable energy potential mainly rely on geographic information system (GIS) data such as solar and wind maps. However, planning systems that realize the potential of renewable energy resources require going beyond GIS tools. In this study, we consider two additional dimensions to GIS maps in the assessment of cities' renewable energy potential: competing space and competing resources.

A. Fall (✉) · R. Haas
Energy Economics Group TU Wien, Vienna, Austria
e-mail: fall@eeg.tuwien.ac.at; haas@eeg.tuwien.ac.at

A. Fall, R. Haas (eds.), *Sustainable Energy Access for Communities*,
https://doi.org/10.1007/978-3-030-68410-5_5

1.1 Study Sample City

The sample city of Dakar overlaps the boundaries of the administrative region, meaning it has a territory of 547 km2 with four administrative subdivisions: Dakar Peninsula (District 1), Guediawaye (District 2), Pikine (District 3), and Rufisque (District 4).

1.2 Renewable Energy Economic and Technical Potential

The concept of competing space includes three layers of technical constraints. The first layer of constraints is the landscape formatted by natural processes over time. The presence of mountains (or valleys), irregular fields (slope), and natural screens such as canopies are all parameters that impose limits to renewable energy resources. At this point, it is worth clarifying the usual confusion between the terms "sustainable" and "renewable" energy resources, with the criteria of landscape co-presence to argue the difference. For instance, it is indisputable to call solar energy a renewable resource, but if the capture of solar irradiation requires to clear the canopy, we consider co-presence to be outraged; thus, this energy cannot be named sustainable. The same goes for scenarios where valleys are levelled to instal hydropower technology.

The second layer of constraints is co-presence in a finite area of the energy technology with human and other animals' settlements and with competing infrastructure. Thus, the assessment of the technical potential excludes areas such as animal parks, birds' reserves, cemeteries, etc.

The third layer of constraints to energy potential realization is the requirement to minimize the distance between the energy transformation system and feed-in-point to the grid in order to increase the energy production potential (EPP) through lowering losses (Masurowski et al., 2016).

The competing resources in the city are conventional energy generation resources that feature the value of having technology infrastructure readily available in the supply mix, which can be numerically accounted as economies of scale. For a technically realizable site to be economically viable, the costs associated with its realization should be lower than that of existing alternatives, which were in majority diesel power generation in the study reference year (2016). The core objective of the chapter is to identify the technical and economic constraints that place boundaries on the infinite theoretical potential of bioenergy, solar photovoltaic and wind energy in our sample city.

2 Methodological Approach

The assessment method is the comparison of unit costs of electricity generation from solar photovoltaic, waste-to-energy and wind energy with the average cost of electricity generation in the study reference year (2016). The cost calculated is the long-run generation cost of electricity. The excel-based computation model uses materials from publications listed below, which are accessible online.

2.1 Assessment of Solar Photovoltaic Potential

The co-presence layer constraints integration of solar photovoltaic systems in the city. The assessment of the city's solar photovoltaic potential relies on second-hand data downloaded from the Global Atlas Map accessible online (Solargis, 2020). From this theoretical potential, we add layers of excluding parameters (e.g. green and water areas), in order to isolate sites where the theoretical potential can be realized with technology, making it the technical potential.

2.2 Assessment of Waste-to-Energy Potential

The distinction between a renewable and a sustainable energy resource excludes from the technical potential of bioenergy, forests and biomass from all areas in the city classified as green or wet. The assessment of the city waste-to-energy potential relies on second-hand data from the city waste collection agency (Unite de Coordination et de Gestion des Dechets, 2016) and from the results of a study on Dakar's waste characterization completed in 2017 as part of a PhD thesis (Fall, 2017). The theoretical energy potential is the potential of energy recovery from quantities of waste collected in the baseline year. From the theoretical potential, we derive the technical potential based on the assumption of a 100% waste collection rate for the city.

2.3 Assessment of Wind Energy Potential

The co-presence with competing infrastructure layer constraints the installation of wind energy technology in the city. The assessment of the city wind energy potential relies on second-hand data downloaded from the Global Wind Atlas accessible online (DTU Wind Energy & ESMAP, 2020). From this theoretical potential, we add layers of excluding parameters (e.g. residences), in order to isolate sites where the theoretical potential can realize with technology, making it the technical potential. Due to our third layer of technical constraint that minimizes the distance between the generation site and grid feed-in-point, we did not consider the wind-offshore potential of Dakar.

The assessment of each of the three technologies economic potential relies on excel-based computation of the long-run generation cost of the technology, which formula is:

Equation 5.1 Long-run generation cost

$$\mathrm{LRGC} = \frac{(\mathrm{CRF\ CAPEX})}{\mathrm{H}} + \mathrm{SRGC}$$

Equation: Long-Run Generation Cost Formula

Where,

$$\mathrm{SRGC} = \frac{F_{\cos t}}{\mathrm{EF}} + \mathrm{Other\ O\&M} + \frac{\left(\mathrm{CO_2}_{\ \mathrm{price}} * \mathrm{CO_2}_{\ \mathrm{fsef}}\right)}{\mathrm{EF}}$$

- CRF is the capital recovery factor.
- CAPEX is the technology capital cost expressed in EUR per kW installed.
- H is full load hours, which are defined as equivalent hours of production at full (rated) capacity. It is the capacity factor expressed in hours.
- F_{cost} is the fuel cost.
- EF is the efficiency factor of the energy conversion process.
- Other O&M stands for other operation and maintenance costs.
- CO_2 price is the price of 1 tonne CO_2-eq emitted by the conversion process.
- CO_{2fsef} is the conversion process-specific emission factor.

3 Results

3.1 Estimation of Solar Photovoltaic Technical Potential

Figure 5.1 downloaded from Solargis shows the city theoretical potential of solar photovoltaic.

Legend: Report generated on 9 January 2020 from https://globalsolaratlas.info/. The energy potential considering the direct normal irradiation (DNI) is 1523 kWh/m2 per year. The surface area of the city is 547 km2. Therefore, the theoretical potential of Dakar on solar photovoltaic is 833,081 GWh per year. Table 1 displays the percent of the land occupied by various uses in the city (Direction de l'Urbanisme et de l'Architecture, 2016).

From Table 5.1, we derive the surface that can host solar photovoltaic systems equivalent to 141.7 km2. Therefore, the technical potential of solar photovoltaic energy is 215,768 GWh. Considering a capacity factor of 17% equivalent to 1489.2 full load hours, this technical potential is 689.9 GW.

Fig. 5.1 Dakar Theoretical solar photovoltaic potential. Source: Global Solar Atlas Website

Table 5.1 Land uses in Dakar (extracted from Urbanism Masterplan towards 2035)

	Land use	
	(in percent)	(in km2)
Dakar		547
Industry and logistics	2.7	14.8
Natural parks	0.1	0.6
Cemeteries	0.1	0.6
Beach and other sand areas	3.4	18.6
Green areas (forests and savannah)	29.6	161.9
Agricultural lands	35	191.4
Wet areas (lakes and rivers)	3.1	16.9
Roads and roadways	0.3	0.6
Other uses (incl. residential and administrative buildings)[a]	29.9	141.7

[a]Other uses group areas that can host solar photovoltaic systems, including shops, administrative buildings, education buildings and nude areas that did not have any geographical or natural affectation

3.2 *Estimation of Waste-to-Energy Technical Potential*

We assess the technical potential of waste-to-energy considering the calorific value of the city waste fractions recoverable as energy. These fractions are organic, plastics, paper, textile and wood waste components. The net calorific value is the approximate net calorific value of common municipal solid waste fractions provided by the World Energy Council (2016), quoting the International Solid Waste Association (ISWA). However, the previous document does not provide calorific value information on the wood component of municipal waste. Therefore, we assume the most common wood species in Dakar, the beefwood tree (Casuarina esquisetifolia), as a source of the wood waste. Information on the calorific value of the beefwood tree (20.9 MJ/kg) is from the World Agroforestry Centre database accessible online. Table 5.2 shows the technical potential of the city waste fractions recoverable as energy.

The technical waste-to-energy potential of Dakar in the baseline year is 39.6 MW power that can be converted with anaerobic digestion (organics) and Dendro Liquid Energy (others) technologies.

3.3 Estimation of Wind Energy Technical Potential

Figure 5.2 downloaded from Global Wind Atlas displays the theoretical potential of wind energy for Dakar.

Legend: Report generated on 10 January 2020 from https://globalwindatlas.info/. The energy potential considering the Mean Power Density of the 10% windiest areas of the city is 320 W per m2. The surface area of the city is 547 km2. Therefore, the theoretical potential of wind energy is 175,040 MW.

As opposed to solar photovoltaic, wind technologies cannot be integrated into the built environment of the city. Wind technologies are ideally installed in peripheral areas of a city to minimize interactions with competing infrastructure such as communication infrastructure. Table 5.1 on soil occupations from the National Directorate in charge of Land Planning (DPU) indicates that about 3.4% of the city's land is made of these nude areas that do not have any geographical or natural affectation (Direction de l'Urbanisme et de l'Architecture, 2016). From this, we can derive a surface area equivalent to 18.6 km2 that can host wind systems with a technical potential for wind energy equivalent to 5951.4 MW.

3.4 Estimation of the Economic Energy Potential

Table 3 lists the input parameters of the Excel model to calculate each technology long-run generation cost in the baseline year (2016) (Table 5.3).

CAPEX is from IRENA (2019) for anaerobic digestion, solar photovoltaic and wind energy. For Dendro liquid energy, CAPEX is from Ghougassian (2012).

Full load hours are derived for each technology from the capacity factor provided by IRENA (2019).

Capital recovery factor (CRF) is computed with a discount rate at 9% (Bah, 2015) and an assumption of 15 years lifetime. CRF is 12.4% for all technologies.

Fuel cost at 0 for all renewable energy technologies is an assumption of the study.

The efficiency factor (EF) is the efficiency of the conversion process. Figures are from Fraunhofer ISE (2019) for solar photovoltaic, and Fraunhofer IWES (Fraunhofer IWES, 2014) for wind energy. For DLE technology, EF is from Ghougassian (2012). For anaerobic digestion, EF is computed with data provided by Wendy et al. (2013), namely:

Table 5.2 Waste-to-energy potential of Dakar

	Fractions (in tonnes)[a]	Net calorific value (MJ/kg)[b]	Energy (GJ)	Energy (MWh)[c]	Full Load Hours (H)[d]	Efficiency factor [e]	Power potential (kW)
Organic	26,058.72	4	104,234.9	29,185.8	7358	0.66	6009.6
Plastics	10,516.56	35	368,079.5	103,062.3	7358	0.8	17,507.6
Paper	5816.68	16	93,066.9	26,058.7	7358	0.8	4426.7
Textiles	7759.45	19	147,429.6	41,280.3	7358	0.8	7012.4
Wood	4653.34	20.9	97,254.9	27,231.4	7358	0.8	4625.9
Glass	3490.01	0					
Metal	2326.67	0					
Other materials	55,712.16	Undefined					
Total			810,065.7	266,818.4			39,584.3

[a]The column compiles the equivalent waste quantities collected in the city in 2016 (Fall, 2017), with an assumption of a 100% collection rate
[b]The net calorific value is from the International Solid Waste Association quoted by the World Energy Council and other publications we accessed
[c]The columns Energy display the results of calculation with previous columns parameters
[d]The column Full load hours assumes 84% capacity factor provided by IRENA (2019) for operating bioenergy power plants in 2016, equivalent to 7358 hours. This figure is within the range (6000–8700 hours) of baseload plants load hours mentioned by IEA Bioenergy (International Energy Agency, 2019)
[e]Source information on figures of the efficiency factor is in Paragraph 4.4

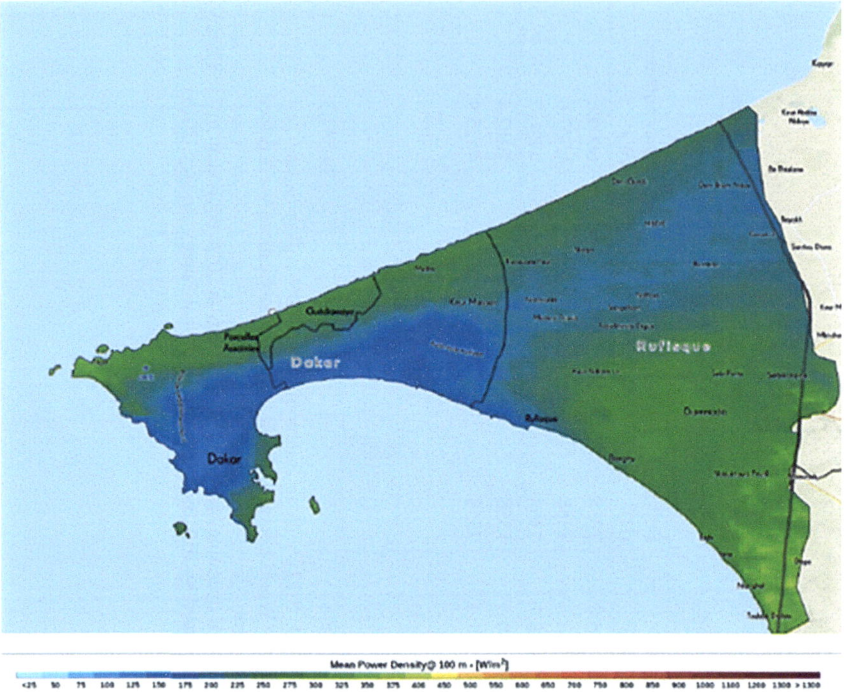

Fig. 5.2 Dakar Theoretical wind energy potential. *Source: Global Wind Atlas*

Table 5.3 LRGC input parameters

Technology	Bioenergy		Solar PV	Onshore Wind
	DLE	AD		
Capacity (MW)	33.6	6	689,945.3	5951.4
Efficiency factor (%)	80	66	21	40
Capacity factor (%)	84	84	17	31
Installation cost (EUR/kW)	517	1890	1609	1609
O&M (EUR/kWh)	0.01	0	0.027	0.01
CO_2 cost (EUR/kWh)	0	0.01		

- Biogas production: 303.6 m3 per ton of municipal solid waste.
- Conversion factor biogas to electricity (in MWh per m^3) is 0.0024.
- Energy content of the organic waste is 1.11 kWh per kg of waste (4 MJ per kg) (World Energy Council, 2016).

Therefore, the efficiency of the anaerobic digestion energy recovery is 65.6%. This rate meets the R1 formula established by the European Commission Waste Directive, which provides that only thermal waste-to-energy conversion with efficiency equal to or higher than 0.65 can be regarded as an energy recovery operation.

Other O&M are from IRENA (2019) for anaerobic digestion, solar photovoltaic and wind energy. For DLE, the other O&M costs are from Ghougassian (2012).

Carbon prices (CO_2 price) available in the literature varies widely per country, and inside the country per sector/project, and per partner buyer. Theretofore, we considered the mean value of the recommendations of the High-Level Commission on Carbon Pricing (World Bank Group, 2019) cited by the United Nations Framework Convention on Climate Change, which is $40–$80 per metric tonne by 2020. Therefore, the carbon price in the study is EUR 54 (60 USD).

CO_2 specific emission factor of the conversion process is estimated at 0 for solar photovoltaic and wind energy. For municipal solid waste-to-energy, the Intergovernmental Panel on Climate Change (2006) estimates that 1 GJ energy conversion of municipal solid waste produces on average:

- 100 kg CO_2 emissions, which are equivalent to 0.36 kg per kWh
- 30 kg CH_4 emissions, which are equivalent to 0.107 kg per kWh. Considering a global warming potential (GWP) of 25, the CO_2 equivalent emissions (CO_2-eq) are 2.7 kg per kWh
- 4 kg N_2O emissions, which are equivalent to 0.014 kg per kWh. Considering a global warming potential of 298, the CO_2 equivalent emissions are 4.29 kg per kWh

Therefore, carbon emissions per kWh generation are estimated at 7.35 kg CO_2-eq.

Conversion rate USD to EUR is 0.9 that was the average rate in 2016 (www.statista.com consulted 7 January 2020).

4 Discussion of Results

4.1 Solar Photovoltaic Potential

The long-run generation cost of solar photovoltaic is EUR cents 16.1 per kWh. This cost is lower than the electricity tariff applied to households (EUR 18.3 cents per kWh), but is higher than the average electricity generation cost (EUR 7 cents per kWh) in the baseline year. Therefore, we confirm an economic potential for decentralized systems installed by citizens who would bear less cost than when using electricity from grid connection. However, at utility scale solar PV is economically not viable in the baseline conditions.

The sensitivity analysis considers the impact on the long-run generation cost of the capital cost (CAPEX) and the discount rate. The unit capital cost of solar photovoltaic can vary with capacity installed due to economies of scale. Large-scale (e.g. utility scale) power plants usually have lower costs than distributed systems. We analyze the sensitivity of LRGC to capital cost in two scenarios and the sensitivity of LRGC to discount rate in two other scenarios.

Fig. 5.3 Solar photovoltaic scenarios comparison

(a) Scenario 1.1: Capital cost is at the level to reach grid parity, all other parameters remaining unchanged. This scenario can also be considered one of utility scale power plants currently in operation, which had a generation cost of around EUR 7 cents in 2016 (SENELEC, 2017).
(b) Scenario 1.2: Capital cost is at the level to reach grid parity. Discount rate is set at 7.5% (OECD average), and all other parameters remain unchanged.
(c) Scenario 2.1: Capital cost is at the level to reach consumer neutrality (i.e. cost = end-user tariff), all other parameters remaining unchanged.
(d) Scenario 2.2: Capital cost is at the level to reach consumer neutrality. Discount rate is set at 7.5% (OECD average), and all other parameters remain unchanged.

Figure 5.3 shows solar photovoltaic capital cost related to scenarios 1 and 2 in comparison to the baseline scenario.

In Scenario 1.1, the capital cost should decrease by 68% to reach grid parity, which is equivalent to a capital cost of EUR 465 per kW installed. In Scenario 1.2, the capital cost should decrease by 65% to reach grid parity with a discount rate of 7.5%, which is equivalent to a capital cost of EUR 509 per kW installed.

In Scenario 2.1, the cost should increase by 13% to reach consumer neutrality over decentralized solar photovoltaic and grid generation, which is equivalent to EUR 1642 per kW installed. In Scenario 2.2, the unit cost should increase by 24% to reach consumer neutrality over decentralized solar photovoltaic and grid generation, which is equivalent to EUR 1798 per kW installed.

4.2 *Waste-to-Energy Potential*

The cost of organic waste conversion to electricity with anaerobic digestion technology (EUR 3.6 cents per kWh) is lower than Dakar's average generation cost in

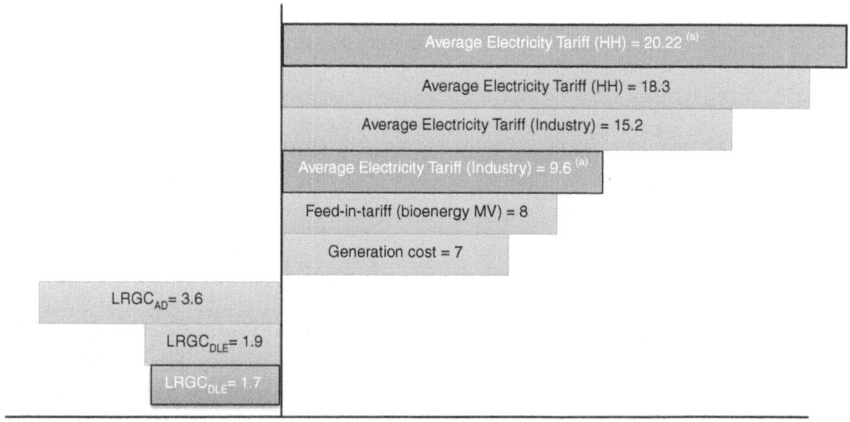

Fig. 5.4 Waste-to-energy costs comparison (EUR cents per kWh)

the study reference year. The cost of other calorific waste materials (plastics, paper, textiles and wood clogs) conversion to electricity with Dendro Liquid Energy technology (EUR 1.9 cents per kWh) is lower than Dakar's average generation cost in the reference year. Therefore, the installation of a 39.6 MW waste to electricity conversion system is economically viable in Dakar. Fig. 5.4 shows the bioenergy generation costs, in comparison to the kWh electricity generation cost, with the grid generation cost in the baseline year as a benchmark.

Compared to the city of Vienna, where similar estimates were completed, discount rate and the quantities of waste collected are parameters that are notably different.

In 2016, Dakar was home to a population of 3.4 million habitants (ANSD, 2020). Fall (2017) estimates the waste generated in Dakar at 171.82 kg per capita per year, which is equivalent to 589,234 tonnes per year. Fig. 5.5a displays a comparison of the energy demand of Dakar and the potential of energy produced by waste at three waste collection quantities, which are:

(a) Scenario 1.1: quantities equivalent to the waste collected in 2016.
(b) Scenario 1.2: quantities equivalent to the total of waste generated (based on Fall estimates), with equivalent proportions of organics and other waste materials.
(c) Scenario 1.3: waste quantities equivalent to that of the city of Vienna (Austrian Federal Ministry of Sustainability and Tourism, 2017), keeping the same materials proportions.

Figure 5b displays a comparison of the energy cost in the baseline year (discount rate = 9%) and the DLE cost at two levels of discount rate, which are:

(a) Scenario 2.1: discount rate is equivalent to 7.5%, which is an average for OECD cities that include Vienna (IRENA, 2019).
(b) Scenario 2.2: discount rate is equivalent to 9% as in the baseline.
(c) Scenario 2.3: Discount rate is equivalent to 14% (fictive).

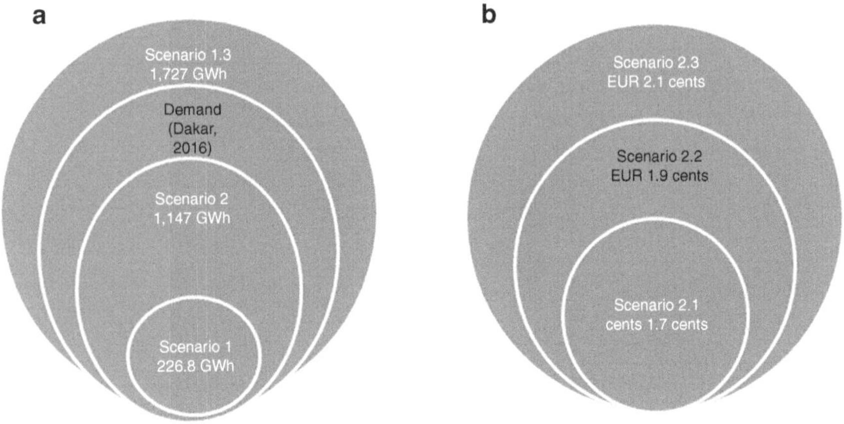

Fig. 5.5 Sensitivity of energy potential (GWh) to waste quantities (tonnes) and capital cost (EUR). (**a**) Simulation of quantities (**b**) Simulation of costs

Scenarios 1.1 and 1.2 (quantities) provide Dakar with the potential to partially cover its energy demand in the baseline year, through recycling its waste, with proportions equivalent to 14% and 68%, respectively. Scenario 1.3 provides Dakar with the potential to cover the overall city electricity demand in 2016. The city's total electricity demand was 1673.6 GWh (ANSD, 2019).

In scenarios 2.1 to 2.3 (capital cost discount rate), the generation cost remains below the generation cost and the tariff in the baseline year. The generation cost (EUR 7 cents per kWh) is relatively high, compared to the costs at different discount rates. The discount rate should be 85% for DLE to reach grid parity, which is unlikely.

Another parameter that is of interest in the waste-to-energy scenarios is CO_2 price. CO_2 pricing assumes a voluntary commitment of Dakar to reduce its carbon emissions as a component of its nationally determined contributions (NDC); Senegal being a non-Annex 1 party to the United Nations Framework Convention on Climate Change (UNFCCC, 2020). We assume the fourth scenario with voluntary carbon pricing that returns a long-run generation cost of:

- EUR 64.1 cents for anaerobic digestion technology
- EUR 51.5 cents per kWh for Dendro liquid energy technology

Therefore, prices for carbon emissions make the waste-to-energy technology cost seven to nine times higher than the average cost of electricity generation, which makes waste-to-energy technology too expensive and no longer a competitive option for Dakar:

4.3 *Wind Energy Potential*

The long-run generation cost of wind energy is EUR 8.3 cents per kWh. This cost is slightly higher than that of Dakar's generation cost for the baseline year. In addition, we assumed that wind energy technology could not be integrated into the built environment, which excludes comparison with end-user tariffs for decentralized generation. Therefore, in the baseline conditions, wind technology is not economically viable. In addition, when we consider the cost as exclusion criteria to compare renewable energy technology with competing energy technologies, the result favours existent generation options (mainly diesel units), as they feature economies of scale accounted in terms of experience on technology use.

The study simulated one sensitivity scenario in wind energy, which sets the discount rate at 7.5%, equivalent to Vienna's figure. The analysis returns that at this rate, the wind cost is EUR 7.7 cents per kWh, which is still higher than the grid generation cost at the baseline. The discount rate should be 5.7% to reach grid parity.

5 Conclusion

This chapter assessed the technical and economic potential of three renewable energy resources available in Dakar, going beyond the geographic information system (GIS) data. The main lesson derived from this assessment is that at local level context-specific parameters like space, technology costs and capital costs matter in making a renewable energy resource economically viable or not. When considering Dakar's space restrictions and the presence of competing (conventional) energy resources, we found that waste-to-energy is the most economical alternative to grid generation considering the reference energy scenario. The technology is readily available, and the city has experience with recycling its waste-to-energy with anaerobic digestion. One example is the Thecogas waste-to-energy plant operational in the main slaughterhouse of Dakar. Dendro Liquid energy is another option for recycling the other calorific waste materials that the city generates in significant quantities (paper, plastics, textiles and wood clogs). Solar photovoltaic and wind energy technologies are also supplying options for the city, as the resources are available and technically usable. However, technology cost and cost of capital (discount rate) should significantly decrease in order for these technologies to be competitive compared to supply from the grid. Still, solar PV can be an alternative to grid, but only at the level of the end-user where tariffs are higher. Wind onshore faces both limitations of space restriction and lack of economic competitiveness, compared to competing resources.

References

ANSD. (2019). *Dakar: Situation economique et sociale regionale*. Agence Regionale des Statististiques et de la Demographie de Dakar. Récupéré sur. https://www.ansd.sn/index.php?option=com_regions&view=region&layout=ses&id=1

ANSD. (2020). *Indicateurs socio-demographiques*. Dakar, Senegal. Consulté le January 25, 2020, sur http://www.ansd.sn/

Austrian Federal Ministry of Sustainability and Tourism. (2017). *Vienna Federal WasteManagement Plan*. Austrian Federal Ministry of Sustainability and Tourism.

Bah, M. (2015). *Détermination d'un taux d'actualisation des projets d'investissement public au Sénégal*. Dakar, Senegal. Consulté le January 21, 2020, sur https://www.researchgate.net/publication/317401014_Determination_d%27un_taux_d%27actualisation_des_projets_d%27investissement_public_au_Senegal

Direction de l'Urbanisme et de l'Architecture. (2016). *Plan d'Urbanisme Dakar Horizon 2035*. Dakar: Direction de l'Urbanisme et de l'Architecture. Consulté le January 3, 2020, sur http://www.pdudakar.gouv.sn

DTU Wind Energy & ESMAP. (2020). *Global Wind Atlas*. Retrieved February 20, 2020, from https://globalwindatlas.info/

Fall, M. (2017). *Caracterisation physicochimique et valorisation energetique des dechetssolides pour la region de Dakar*. EPT.

Fraunhofer, I. W. E. S. (2014). *Windmonitor - share of different turbine capacities on thecumulative*. Fraunhofer IWES. Retrieved January 21, 2020, from http://www.windmonitor.de/windmonitor_en/bilder_javascript.html?db_communicate=%27Windenergieeinspeisung.daten%27&p_lang=eng&img_id=433

Fraunhofer Institute for Solar Energy Systems. (2019). *Photovoltaics report*. Fraunhofer Institute for Solar Energy Systems. Retrieved January 23, 2020, from www.ise.fraunhofer.de

Ghougassian, B. (2012, November 26). *Waste To Energy Technologies*. Retrieved January 21, 2020, from Arab Forum for Environment and Development: http://www.afedmag.com/english/ArticlesDetails.aspx?id=12

Intergovernmental Panel on Climate Change. (2006). *IPCC Guidelinesfor National Greenhouse gas Inventories Volume 2: Energy*. Intergovernmental Panel on Climate Change.

International Energy Agency. (2019). *The future roleof thermal biomass power in renewable energy systems – A study of Germany*. IEA Bioenergy.

IRENA. (2019). *Renewable power generation cost in 2018*. International Renewable Energy Agency.

Masurowski, F., Drechsler, M., & Frank, K. (2016). A spatially explicit assessment of the wind energy potential in response to an increased distance between wind turbines and settlements in Germany. *Energy Policy, 97*, 343–350. https://doi.org/10.1016/j.enpol.2016.07.021

Ng, W. P., Varbanov, P. S., Klemeš, J. J., Hegyháti, M., Bertók, B., Heckl, I., & Lam, H. L. (2013). Waste to energy for small cities: Economics versus carbon footprint. *Chemical Engineering Transactions, 35*. https://doi.org/10.3303/CET1335148

SENELEC. (2017). *Rapport annuel*. SENELEC.

Solargis. (2020). *Global Solar Atlas*. Retrieved January 21, 2020, from https://globalsolaratlas.info/map

UNFCCC. (2020, April 18). Parties to the convention. Retrieved from United Nations Climate Change: https://unfccc.int/process/parties-non-party-stakeholders/parties-convention-and-observer-states

Unite de Coordination et de Gestion des Dechets. (2016). *Programmede Gestion des dechets solides urbains de la region de Dakar*. UCG.

Wolsink, M. (2018). Co-production in distributed generation: Renewable energy and creating space for fitting infrastructure within landscapes. *Landscape Research, 43*(4), 542–561. https://doi.org/10.1080/01426397.2017.1358360

World Bank Group. (2019). *Report of theHigh-level commission on carbon pricing and competitiveness*. Washington D.C. Retrieved February 2020, from https://openknowledge.worldbank.org/handle/10986/32419

World Energy Council. (2016). *World energy resources. Waste-to-Energy*. World Energy Council. Retrieved January 23, 2019, from https://www.worldenergy.org/assets/images/imported/2016/10/World-Energy-Resources-Full-report-2016.10.03.pdf

Chapter 6
Energy Potential of Crop Residues in Senegal: Technology Solutions for Valorization

C. A. Mbodji, A. Fall, D. Diouf, and A. Seck

Abstract Different crops are harvested in Senegal at different periods throughout the year. Therefore, crop residues are constantly available, and there is potential for recycling these residues to produce energy. However, most of these crop residues are often burned or leftover in fields. This chapter assesses the energy potential of five crop residues available in Senegal, in terms of raw material quantities and calorific values, in order to propose adequate valorization schemes. The methodology of the study is based on a review of scientific literature in the field, the processing of data collected from the national specialized agencies, and laboratory tests of pellets manufactured from these residues. The experiments also propose techniques of hybridization in pelleting. The results show the feasibility of combining crop residues to produce pellets, with improved energy characteristics. They also demonstrate that pellets from crop residues can substitute part of industries' heat demand with the combustion technology and part of rural communities' electricity demand with the gasification technology.

Keywords Crop residues · Pellets · Biofuels · Gasification · Combustion

---

C. A. Mbodji (✉) · D. Diouf
Laboratoire Electronique, Informatique, Télécommunications et Energies Renouvelables, Gaston Berger University, Saint-Louis, Senegal
e-mail: mbodji.cheikh-ahmadou@ugb.edu.sn; djicknoum.diouf@ugb.edu.sn

A. Fall
Energy Economics Group TU Wien, Vienna, Austria
e-mail: fall@eeg.tuwien.ac.at

A. Seck
SEnergyS Africa Company, Dakar, Senegal
e-mail: abdoulayeseck@senergysafrica.com

A. Fall, R. Haas (eds.), *Sustainable Energy Access for Communities*,
https://doi.org/10.1007/978-3-030-68410-5_6

1 Introduction

Senegalese agriculture mainly relies on subsistence crops. This agriculture has long been seasonal, but recent initiatives brought in mechanization and harvesting periods out of traditional seasons, as well as crop diversification.

In its Intended Nationally Determined Contribution (INDC), Senegal ambitions to produce 4,000,000 Gigajoules (685,000 MWh) of biofuels annually, recovered from waste (Ministere en charge de l'Environnement, 2015). The recovery process would avoid 457,000 tonnes of CO2 per year. This objective constitutes the framework of current initiatives using crop residues to produce fuels for industrial applications. Our study specifically targets the use of these residues for producing pellets.

The requirements of the pelleting process are specific to each crop residue and depend on its characteristics. The parameters that we consider meaningful in the process are the physical characteristics of the biomass (e.g. moisture content), the energy output required (e.g. gas or electricity), and the economic conditions throughout the value chain from the collection of residues in fields, followed by transport and preparation in situ of raw materials (e.g drying and grinding) to the production of pellets. The chapter provides valorization schemes that convert five crop residues into pellets. Our methodology combines in-laboratory experiments where the characteristics of crop residues are measured, and different hybridization schemes are tested, followed by documentation of results from the experiments. We confront these results with lessons learned from previous studies on pelleting and the use of pellets. Before, we start by a review of the literature on the state-of-the art of crop residues production in Senegal.

2 Potential of Crop Residues in Senegal

The crop residues in our study sample are residues that have a potential for energy recovery and are available in significant quantities in different regions of the country. These include groundnut shells, palm nutshells, corn cob, rice husk, and bulrush. Figure 6.1 shows these residues, with quantities available per year and harvest areas.

Groundnuts (*Arachis hypogaea*) are harvested mainly in the so-called groundnut basin (centre of Senegal). The potential of groundnut shells available in this area is estimated at 142,000 tonnes (ANSD, 2019). The groundnut shell, despite its light weight, has a density of 270 kg/m3. Its low calorific value (LCV) is 16,704 kJ/kg (4.64 kWh/kg) (BRISK-2-Project, 2019).

Palm nuts (*Elaeis guineensis*) are produced mainly in the Lower Casamance region (south of Senegal). The potential of palm kernel shells available in this area is estimated at 50,000 tonnes (ANSD, 2019). The palm kernel shell features both the highest density and calorific value of the five crop residues considered, with

Fig. 6.1 Crop residues potential and location in Senegal

630 kg/m3 and 25,095.6 kJ/kg equivalent to 6.971 kWh/kg (BRISK-2-Project, 2019).

Maize (*Zea mays*) is harvested mainly in the centre, and south-east regions of Senegal. The corn cob potential of this area is estimated at 348,000 tonnes (ANSD, 2019). The corn cob has a density of 270 kg/m3 (BRISK-2-Project, 2019) and a low calorific value (LCV) of 24,760.8 kJ/kg (6.878 kWh/kg).

Rice (*Oryza glaberrima*) is traditionally cultivated during the rainy season in southern Senegal (Casamance) and now out-of-season in northern Senegal. The rice husk potential from both regions is estimated at 260,000 tonnes (ANSD, 2019). The rice husk has a low density of 120 kg/m3 (BRISK-2-Project, 2019) and a low calorific value (LCV) of 14,004 kJ/kg (3.89 kWh/kg).

Bulrush (*Typha australis*) is an aquatic plant that grows along the Senegal River. Bulrush cannot be directly linked to a food crop, unlike the other residues cited previously, but was included in this study considering its availability and its energy potential. The quantities of bulrush produced annually are estimated at 520,000 tonnes of dry material (ANSD, 2019). The bulrush has a significant energy potential, with a low calorific value (LCV) of 17,388 kJ/kg (4.83 kWh/kg) and a density of 120 kg/m3 (BRISK-2-Project, 2019). Table 6.1 summarizes the quantities of crops and their residues in the study sample.

Table 6.1 Estimates of crops production and residues in Senegal in 2018 (ANSD)

Crop	Production (tonnes)	Residue	Production (tonnes)
Groundnut	1,432,086	Groundnut shells	142,000
Palm	71,500	Palm kernel shells	50,000
Corn	476,600	Corn cob	320,000
Rice	1,132,800	Rice husk	260,000
Bulrush	3,050,000	Dry bulrush	520,000

ANSD (National Agency for Statistics & Demography) is a public agency in charge of periodic publications on statistics in the agriculture sector (ANSD, 2019)

Table 6.2 Characteristics of experimental pellets

Pellets produced	Humidity level	Ash rate	Volatile content	Fixed carbon	LCV (kJ/kg)
Bulrush + palm kernel	6.94%	7.48%	89.34%	3.18%	18,549
Groundnut shell + palm kernel shell	8.63%	3.32%	77.99%	18.69%	19,258
Bulrush + corn cob	5.68%	7.97%	90.59%	1.44%	17,687

3 Pellets Processing

The processing of crop residues through pelleting provides a wide range of energy recovery schemes. The pelleting protocol remains the same for the five crops' residues in our sample. A preliminary drying phase is necessary, followed by mechanic grinding. After grinding, the residues are compacted and transformed into pellets. The major challenge in pelleting is the moisture content before compaction. The challenge is more important in the hybridization scenarios, where we combine biomass materials with different properties. The experimental pelleting process we conducted required a hammer mill and a pelletizer with a power of 7.5 kW each.

The experiment returned high-quality and high-energy pellets. The physical characteristics of pellets produced were tested in the laboratory. Table 6.2 shows the physical characteristics of the pellets, which are produced through mixing different crop residues.

The physical properties of the pellets in Table 6.2 make them suitable for use in the thermal processes of various industrial applications; besides the energy potential, they have low humidity and low ash rates. The combination groundnut shell and palm kernel shell returns the highest calorific value, because the two residues have a high percentage of fixed carbon, which made the mix easy to homogenize.

4 Technology Solutions for Valorization of Pellets

Pellets previously produced can be used in both thermal and electrical applications. In thermal applications, the pellets are used directly as fuel in industrial boilers and ovens to generate. In the electrical applications, the process requires to convert the pellet energy content into steam that powers electricity production.

4.1 Combustion

The combustion of biomass pellets produces hot gases at temperatures around 800–1000 °C (McKendry, 2002a, b). In our first experiment, the pellets that result from the combination of groundnut shells and palm kernel shells were tested in a bakery oven as fuel to produce the heat needed for baking bread. This experiment showed that the pellets were more efficient in terms of quantities and energy generated than the conventional fuel (charcoal) used in the bakery.

The five crop residues cited in this article, with the exception of risk husks, could also be burned directly in waste-to-energy plants without any chemical processing to produce steam for generating electricity. In this case, fluidized bed combustion would be the best combustion technology (Saidur et al., 2011). Fluidized bed combustion has emerged as a viable alternative to conventional combustion systems and offers multiple benefits such as compact boiler design, fuel flexibility, higher combustion efficiency, and reduced emission of noxious pollutants. The fluidized bed boilers have a wide range of capacities, from 0.5 T/h (tonnes/hour) to over 100 T/h (Saidur et al., 2011).

4.2 Gasification

The gasification process converts pellets into a combustible gas mixture (syngas) by partial oxidation of the pellet at high temperatures (800–1500 °C) (McKendry, 2002a, b).

The gasification process can be used for direct gas production, but also for electricity production (conversion of the gaseous fuel to electricity through gas turbine). The quantities of rice husk, groundnut shell, palm kernel shell, bulrush, and corn cob have an energy potential of 7400 GWh/year. This potential can fuel large power plants in the areas where the crop residues are produced, which operate according to a proven protocol: gasification of the residues and production of electricity with an alternator coupled to a combustion turbine. The average efficiency of such systems is estimated at 35% (McKendry, Energy production from biomass (part 3): gasification technologies, 2002a, b). The option is particularly interesting to consider given areas in Fig. 6.1 where the residues are produced are yet to achieve universal access to electricity.

Our second experiment consisted of using the pellets in a cooking stove. The results showed that the stove with pellets was as efficient as liquefied petroleum gas (LPG) cooking stoves in terms of cooking time. Our estimates showed that by using the pellets in Table 6.2, a household in Senegal could save up to 50% in its cooking energy bill with a better energy service.

4.3 Anaerobic Digestion

The anaerobic digestion converts the crop residues into biogas, which is a mixture of methane and carbon dioxide, with small quantities of other gases such as hydrogen sulphide (McKendry, 2002a, b). The biogas produced has an energy content equivalent to about 20–40% of the lower heating value of the feedstock. Anaerobic digestion and biochemical conversion, in general, are usually preferred for biomass with high moisture content such as manure.

In Senegal, the National Domestic Biogas Programme (PNB-SN) was developed in order to promote this technology in rural areas, with cow and donkey dungs as fuels. Recent developments in technology opened the possibility of using paddy rice husks and other crop residues in biogas technologies with the solid-state anaerobic digestion (SS-AD) digesters. Indeed, SS-AD operates with biomass fuels that have a proportion of solid content higher than 15%, making it particularly suitable for digesting the lignocellulose content of rice husks (Matin & Hadiyanto, 2018). The rice husk potential (260,000 tonnes per year) makes SS-AD relevant for biogas production in northern and southern Senegal, therefore providing an eco-friendly solution to the management of waste, which has so far been difficult to value due to material composition.

5 Conclusion

Senegal has a considerable potential for generating electricity from crop residues, which is not fully valued as of today. In this chapter, groundnut shells, palm kernel shells, corn cobs, rice husks and bulrush were exemplified as residues with high energy potential and high material quantities.

The research experiments build from previous studies that demonstrate the energy potential of these crops taken individually and introduces hybridization schemes that benefit from the complementary characteristics of two or more crops combined into pellets. Our results open new perspectives on the design of pellet mixtures that suit different energy recovery techniques. Our results show that the most promising options for energy recovery from pellets are combustion and gasification. Our experiments also demonstrate the potential of pellets' use in small-scale businesses such as bakeries and restaurants. Besides, the pellets could also replace conventional fuels in ovens and boilers in industrial applications. In fact, our pellets feature the physical characteristics of high calorific values and low moisture content and ash rate, which make them reliable combustion fuels.

Future studies could delve into heat generation with pellets in a combustion technology and test the potential of various types of pellet mixtures to substitute fossil fuels in industrial boilers. It would also be interesting to explore the economic potential of pellets for electricity generation in small- and medium-scale transformation companies that could add value to crops initially harvested.

Acknowledgments The authors acknowledge the support of SEnergyS Africa Co. for providing the materials used in the experimental studies and the production of pellets.

The authors acknowledge the support of the Centre for Studies and Research on Renewable Energy of Dakar (CERER) for providing laboratory facilities to test the characteristics of our pellets.

References

ANSD. (2019). *Bulletin mensuel des statistiques economiques*. Agence Nationale des Statistiques et de la Demographie.

BRISK-2-Project. (2019). *Database for the physico-chemical composition of (treated) lignocellulosic biomass, micro- and macroalgae, various feedstocks for biogas production and biochar*. Stockholm. Retrieved August 2, 2019, from https://phyllis.nl/

Matin, H. H., & Hadiyanto, H. (2018). Biogas Production from Rice Husk Waste by using Solid State Anaerobic Digestion (SSAD) Method. *E3S Web of Conferences. 31*, p. 7. EDP Sciences. https://doi.org/10.1051/e3sconf/20183102007.

McKendry, P. (2002a, May). Energy production from biomass (part 2): Conversion technologies. *Bioresource Technology, 83*(1), 47–54. https://doi.org/10.1016/S0960-8524(01)00119-5

McKendry, P. (2002b). Energy production from biomass (part 3): Gasification technologies. *Bioresource Technology*, 55–63. https://doi.org/10.1016/S0960-8524(01)00120-1

Ministere en charge Environnement. (2015). *Contribution prevue determinee au niveau national*. United Nations Framework Convention on Climate Change. Consulté le August 2, 2019, sur https://www4.unfccc.int/sites/submissions/INDC/Published%20Documents/Senegal/1/CPDN%20-%20S%C3%A9n%C3%A9gal.pdf

Saidur, R., Abdelaziz, A., Demirbas, A., Hossain, M. S., & Mekhilef, S. (2011, June). A review on biomass as a fuel for boilers. *Renewable and Sustainable Energy Reviews, 15*(5), 2262–2289. https://doi.org/10.1016/j.rser.2011.02.015

Chapter 7
Democratization of Energy Planning: On a New Planning Tool Tailored to the Needs of Developing Countries

A. Fall, C. H. T. C. Ndiaye, and C. A. Mbodji

Abstract The chapter presents the concept of a model developed to support planning of energy access and transition to sustainability in emerging cities of the sub-Saharan Africa region. Energy planning software is the most used tool for the description and planning of an energy system. Existing software features several drawbacks that make them irrelevant for planning energy systems in cities like Dakar, Senegal. This chapter documents the reasons why existing energy planning software is not suitable for planning these energy systems. It also proposes a new approach for designing a software (MoCES) that is tailored to the needs of Dakar, using a combination of multi-criteria decision-making analysis, energy landscaping, and visual computing. MoCES is an innovative tool that supports the rising citizens' engagement to access and transition to a sustainable energy future, which integrates all dimensions of sustainability, including economic affordability, social acceptance and environment friendliness.

Keywords Energy Sustainability · Planning software · Cities · Sub-Saharan Africa

The word emerging for the city of Dakar considers classification of the World Cities by the New Climate Economy Report (2014), meaning "rapidly expanding middle-income, mid-sized cities (…), with populations of 1–10 million, and per capita incomes of US$2000-20,000."

A. Fall (✉)
TU Wien Energy Economics Group, Vienna, Austria
e-mail: fall@eeg.tuwien.ac.at

C. H. T. C. Ndiaye
Universite Cheikh Anta Diop de Dakar/Ecole Polytechnique de Thiès, Dakar, Senegal

C. A. Mbodji
Laboratoire Electronique, Informatique, Télécommunications et Energies Renouvelables, Gaston Berger University, Saint-Louis, Senegal

A. Fall, R. Haas (eds.), *Sustainable Energy Access for Communities*,
https://doi.org/10.1007/978-3-030-68410-5_7

1 Introduction

There is a momentum for individual actions in the promotion of energy sustainability in cities. Grubler et al. (2012) estimate that about 80 percent of all commercial energy used in the sub-Saharan Africa region can be classified as urban. This urban energy is a major contributor to greenhouse gas emissions in the region. However, the planning of energy systems remains largely a competence of national governments. Ringkjøb et al. (2018) identified 75 tools developed in different regions of the world to model planning, operation, and monitoring of energy systems. Majority of these tools model the energy systems at national and regional (group of countries) levels. Among the planning software listed by Ringkjøb et al. (2018), we tried to select a software for modelling the Dakar energy system in a 3-step filtering method based on the following criteria:

(a) Experience of use in a developing country returns 40 models (Urban et al., 2007).
(b) Focus on energy and electricity and exclude models that entirely focus on climate change and its impact or address economic issues such as energy markets returns 12 models (Urban et al., 2007).
(c) Possibility to adapt the model to a scale below the national level (city) returns 0 models.

The reasons why existent energy planning models are not relevant for Dakar relate to limits on the modelling approach, the energy accounting methodology, and the empirical validity of the models' outcomes.

1.1 Limits of the Approach

For Van Beeck (1999), the implicit assumption of constancy from historical trends in energy models with a top-down approach is not realistic, because the rapid population and economic growth, which characterize cities emerging cities, affect energy use. For Urban et al., the bottom-up approach weaknesses relate to the non-tractability of parameters such as demand, technology change, and resources assumed in the model as exogenous while being main drivers of the system's dynamics (Urban et al., 2007).

1.2 Limits of the Methodology

The adaptation to the city level of energy planning software originally developed for national and regional contexts requires convergence criteria and linking metrics, meaning energy metrics that link the national and the city scales. Grubler et al.

(2012) identify four methods for accounting the city energy use, which are further classified as production-oriented methods (2), and consumption-oriented methods.

The two production-oriented methods require to fix the city's boundaries. Barles (2009) illustrates the variability of energy metrics depending on whether boundaries are set to Paris, extended to its suburbs or include the larger Parisian metropolitan region.

The two consumption-oriented methods require to define an economic metric at the city level, which is usually the gross regional product, to link with a national metric that can be the gross domestic product. Both metrics exclude the informal economy, whose proportions are not linear. The share of the informal economy can be relatively important in the region; it was estimated at an average 40.2% of the total official GDP in sub-Saharan Africa during the period 1999–2007. Comparatively, the share was 17.1% in OECD countries (Schneider, 2012).

1.3 Limits of Validity

The existing models lack some context-specific characteristics of energy systems in developing countries, which affect the empirical validity of their outcomes. The models' assumptions are usually biased by experience from the energy systems of industrialized countries where they are tested. Those biases include the assumption that the energy systems of developing countries would behave like those of industrialized countries (Shukla, 1995), which explains the absence of parameters related to access, structural economic changes (e.g. urbanization), the predominance of bioenergy, etc.

This chapter aims to document the architecture of existing models and the architecture of an innovative approach in modelling the dynamics of a city energy system, which learned from the experiences gathered on modelling complex systems in computer sciences and in economics. It proposes a new energy planning model that should better support the city of Dakar in planning a transition to energy sustainability.

2 Architecture of Existing Energy System Planning Software

2.1 Modelling Approach

Since the 1970s, following the first oil crisis, energy planning software has been developed to: (i) model the interdependence between the energy system and the general economy, (ii) reduce countries' dependence on fuel imports, and (iii) simulate the environmental impact of an energy strategy. Different criteria can be considered for classifying the modelling approach of these software. Here, we consider with Van Beeck (1999) a distinction based on the approach that classifies existing energy models into three categories.

- Models with a top-down approach considered as macro-economic modelling use a methodology based on the computation of the system equilibrium. This category features ETA-MACRO and Computational General Equilibrium (CGE) models.
- Models with a bottom up-approach considered as micro-economic modelling use a methodology based on optimization of one of the system parameters (e.g. cost). This category features MARKAL and MESSAGE.
- Models with a hybrid approach combine the two approaches above described.

In the bottom-up approach, where models describe an open system, Schrattenholzer (1981) further identifies three (3) sub-categories of models based on dynamics of the system described.

- Models formulating laws of nature that are valid in all circumstances (e.g. the motion of a mass point in a gravity field: $|F| = G * (m1*m2)/r2$).
- Models formulating regular behaviour that have a localized validity (e.g. energy generation in relation to operation hours: $E = \alpha H$).
- Models formulating concepts of controlled man-made systems that provide a quantitative conceptualization of a complex system. The majority of existing energy planning models fit in this category.

2.2 Assumptions and Data Organizing

Among the many energy planning models, the computational equilibrium model is particularly frequent in the literature describing the Senegal energy system (Benedict, 2013). The computational equilibrium model, by its nature of a mathematical based tool, returns relevant outcomes in a stable market environment that fulfils pre-conditions of the neo-classical economics principles, meaning symmetry of information and rationality of agents who are perfectly capable of anticipating and optimizing choices that should drive the system to equilibrium. The logarithm function introduces randomness in the stochastic version of the model but still fails to capture boom-and-bust cycles that happen in real systems. Indeed, the rationale of mathematical theorems is to provide the same results when they are fed with the same data.

Case Study of MESSAGE Application to the City of Vienna (Messner & Strubegger, 1995)

Optimization is another common approach in planning the energy system of cities. The application of MESSAGE to the context of Vienna in Austria sets an optimization function with three objectives:

- Minimum energy cost,
- Minimum fuel imports, and
- Minimum level of indexed pollutants.

Each objective is formulated as a target trajectory. Then, the software computation consists in trying to equally reach each of the target trajectories. The optimal points derive from the optimization of each time step of each objective. The optimal points form the utopia trajectories. These trajectories are utopic in the sense that, although each point can be reached individually, there is no solution that reaches all of them. The worst solution for each time step and objective forms the nadir points. From these points, the software derives the shape of the Pareto-optimal border of the solution space. A mixed-integer-programming (MIP) model represents all options in the solution space as integer variables. From this process, Messner and Strubegger projected the optimal energy mix of Vienna by 2015 (20 years later).

Empirical Relevance of Models' Outcomes

The two models exemplified above (CGE and MESSAGE) perform correctly in environments with characteristics similar to assumptions embodied in the software, i.e. cities with large energy supply technologies, which assume continuous model variables. Outside these environments, manipulations to adjust technologies' sizes or convert continuous variables to discrete values through mixed-integer programming affect the outcomes and their relevance to predict the real system. In addition, the linear transfer of the optimization models' algorithms from their primary field of application, which is conventional energy systems, to renewable energy systems is another point of concern, because only two of the many dimensions of sustainability are considered in the model: economy and environment. Still, this consideration is unrealistically restrictive, with only cost and carbon dioxide (CO_2) metrics tracked in the model.

3 Architecture of an Innovative Energy Planning Software

3.1 Modelling Approach

Cities are identified in the urban studies' literature as complex and self-organizing systems (Batty, 2005). Therefore, to escape the routine of approaching the energy planning model from either the top or the bottom, we propose a model based on the rules of the city agents. For instance, the decision of an agent (e.g a small transformation industry) without energy access either to connect to the grid or to consider another supply option, for example solar photovoltaic, can be from a basic rule: the energy value of the solar photovoltaic generation should be higher than the value of electricity from the grid for the agent to access with solar photovoltaic. If the value of electricity from the grid happens to be higher, then the agent accesses with the grid. The value in this example can be conceptualized as the inverse of the cost of 1 kWh, assuming consumer neutrality over the electricity production technologies.

3.2 The Entropy Dimension in Modelling Complex Systems

For Hayek (1948), the order in a market system is necessarily emergent (spontaneous) and cannot be the result of a central planning. Another element of system entropy is inferred from the second law of thermodynamics, which states that any local change of order bears a cost (Georgescu-Roegen, 1971). This cost can be labelled as energy in the case of physics applications or as greenhouse gas emissions in the case of an energy system. Failing to account for this entropy law is another cause of equilibrium and optimization models' irrelevance for our sample city. The agent taken as example in the previous paragraph, with a basic rule on the energy value, keeps the system order emergent, but makes the potential to have a low-carbon electricity generation a probability measure that depends on the energy value rather than an optimum figure derived from sophisticated computations or a hypothetical equilibrium. The innovative software can weigh this value considering different attributes associated with electricity production technologies.

3.3 Capturing the Entropy Value in MoCES

The main principle that guided the development of an innovative energy modelling software named MoCES (modelling cities energy systems) is inspired by the Aristotle principle of system theory stated in the Metaphysics: the whole is more (or less) than the sum of its parts. Bertalanffy (1969) proposes two approaches for modelling a system: (1) to define general laws that are fit to empirical observations; and (2) to arrange the empirical fields in a hierarchy of complexity of the basic individual behaviours. Computational equilibrium models use the first approach, which can be considered with Bertalanffy as valid, "but makes unreasonable simplifying assumptions" according to Focardi (2015).

MoCES captures the complexity of agents in the system by providing room for a wide range of rationales that can motivate the system planner beyond costs and CO_2 emissions. Examples of these motives include:

- Anticipation of the future: the agent foresees gain losses in the future due to a policy scenario. For instance, investment in a rooftop solar photovoltaic system at T could be more or less advantageous than making the same investment at T + 1 when the regulator could revise the incentive policy (e.g. feed-in-tariff). A relevant metric would be the net present value (NPV).
- Conscious planning: the previous agent could have a preference on gradual independence from the utility than on saving in the capital investment. A relevant metric would be the payback period.
- Dedication to a cause: the same agent could have a preference on protecting the immediate and/or distant network and would prioritize reduction of greenhouse gas emissions over saving money on technology investment or electricity bill. Relevant metrics include carbon dioxide and other greenhouse gas emissions (e.g. methane and nitrous oxide).

Table 7.1 Comparison of computational equilibrium and agent-based modelling for energy systems

		Computational equilibrium	**Agent-based simulation**
Criteria		1	Unlimited
Information		Benchmark	Forecast
Outcome relevance		General	Context-specific
Sustainability dimensions	Economy	Yes	Yes
	Environment	No	Yes
	Society	No	Yes

Table 7.1 displays a comparison of computational equilibrium models and multi-criteria agent-based models considering three modelling criteria.

4 The Modelling Energy System Software (MoCES)

4.1 Data Organizing

The energy planning software we developed includes two main components that are simulation and visualization windows.

The simulation window is common in the energy planning software interface. MoCES simulation window computes various parameters of an energy system, taking into account the multi-faceted dimensions of sustainability, including economic affordability, social acceptance and environment friendliness. MoCES has the capacity to compute ninety-three (93) outcomes based on the agent's inputs.

The visualization window in planning energy systems is an innovation that MoCES brings. It builds from the most recent developments in visual computing and 3D modelling in information technology. The window models animation and 3D rendering of a virtual energy system installed by the user. MoCES visualization window uses geolocation to view the space where the energy system is planned, with three levels of zoom from city to building, and the district level (see Fig. 7.1).

Both windows are connected, which enable the software to automatically adjust the virtual energy system when the simulation data change.

4.2 Programming Interface

Interactive and static plans of MoCES are integrated with Google Maps API. The combination of Street view and high-resolution satellite images returns a good rendering that is precise enough to enable the user to delineate the space dedicated to the energy system.

Graphic materials displayed in 3D and 2D are created with WebGL whose functionalities include liaison with the hardware and the software system accessible through the internet connection.

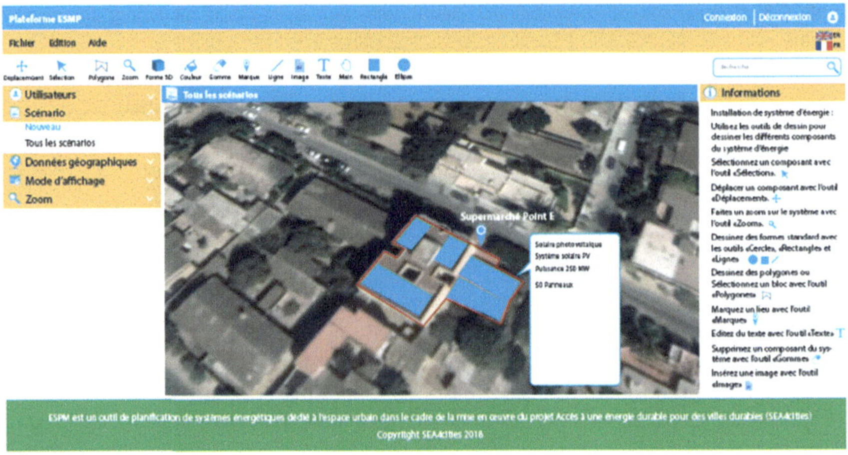

Fig. 7.1 MoCES delineation of the energy system target space

For the programming interactions in the environment, MoCES uses Unity 3D of Unity Technologies, which functionalities (assets) make it possible to import a large variety of image and audio applications that are compatible with various media..

4.3 *Data Management and Security*

The MoCES application is developed under PHP, HTML5, CSS, and Javascript. Access to the application is through the internet connection, which requires to secure the data of prospective users. For the first contact of the user with the application, we set up an SSL protocol to access the platform, which secures authentication and avoids network sniffing. For securing the data entered in the simulation window, in addition to the input control, the application features prepared queries via PDO. Those appear to be efficient against possible security breaches such as injection SQL and provide cleaning services. Files are transferred through the FTP protocol coupled with SSL to authenticate the user certificate during connection.

The implementation process organizes data recorded in the application in three databases: (1) geographic data, (2) energy data, and (3) users' data. The system administrator is able to update all data, confirm users' authorization to connect and prevent possible abuses. Figure 7.2 shows MoCES data storage and management techniques.

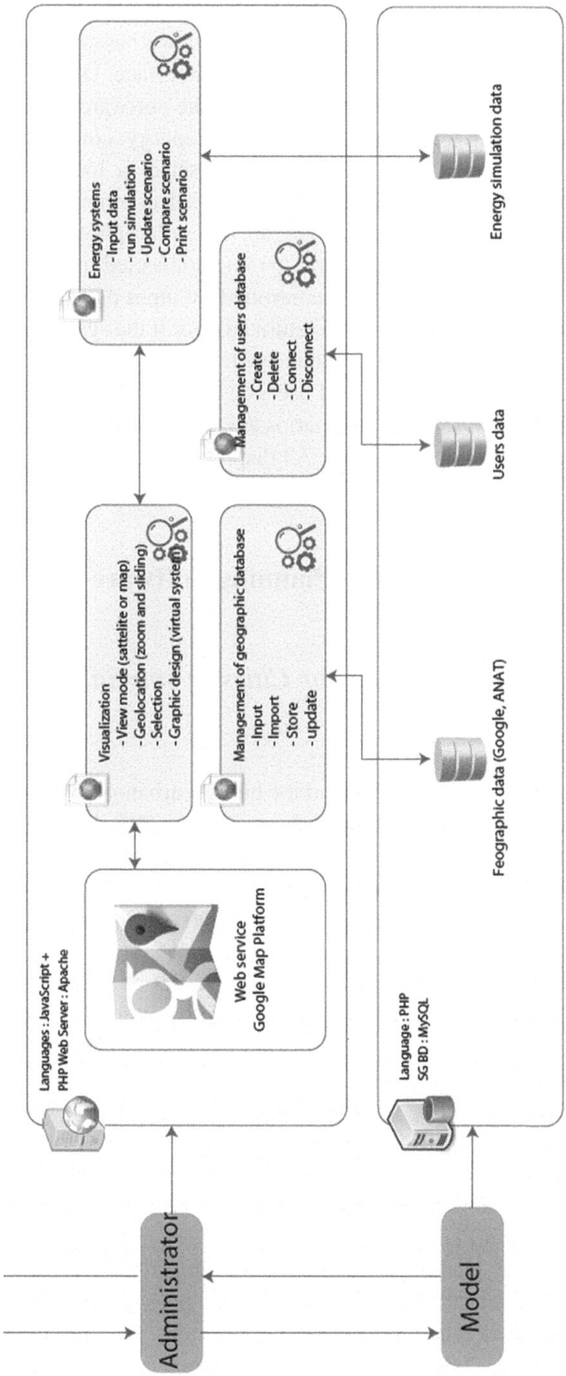

Fig. 7.2 MoCES data management

4.4 Reproducibility of Model Outcomes

The user interface enables the creation of energy scenarios based on the users' input data and default data embedded in the programming interface. Default inputs aim to support non-experienced energy planners in using the software with data that are easily accessible to them (e.g. building position, energy consumption periods, energy supply options, etc.). The user is able to select a location on the map (industrial commercial, or residential building), and select among the energy technologies embedded in the software catalogue: (1) solar photovoltaic, (2) solar thermal, (3) wind, (4) waste recycling, and (5) interconnected grid. The size of the energy system and its capacity are back-controlled by input data already entered in the scenario. For instance, the system will return an error if the system requirement in terms of size is higher than the surface delineated, including the necessary distance between PV panels or wind turbines. All users can create scenarios and store these scenarios as public or private. Public scenarios are accessible to other users who can modify, export and/or print them. Figure 7.3 displays MoCES results window.

5 MoCES and Other Energy Planning Software in sub-Saharan Africa

5.1 Planning Energy Systems for Cities in sub-Saharan Africa

Over the last two decades, Africa recorded the highest urban growth, with the urban population projected to increase from 36% in 2010 to 50% by 2030 (Transform Africa, 2017). The capital city of Senegal, Dakar, is an illustration of this dynamic, with a population that increased from 400,000 in 1970 to 3.4 million inhabitants in 2016. Therefore, cities in the region are the next frontier in assessing the region's capacity to cope with the major challenges of a sustainable development that integrates urbanization and climate variability. Cities of the region also feature another peculiar, which is that they are yet to reach the goal of universal access to sustainable energy (SDG-7). Scientific publications on energy planning software in sub-Saharan Africa are often studies that compare conventional energy (e.g oil) and renewable technologies (e.g solar photovoltaic) (Trotter et al., 2017). In many, the modelling approach consists on the optimization of parameters of the energy generation system such as size, cost, or level of greenhouse gas emissions. Examples of software developed for the region includes an energy planning software for Nigeria and the Network Planner.

In Nigeria, an energy planning software was developed to select among technology options to access electricity. The software compute optimized electrification pathways and optimized renewable energy applications for improving electricity access (Bertheau et al., 2016). The software approach is: (1) to identify energy

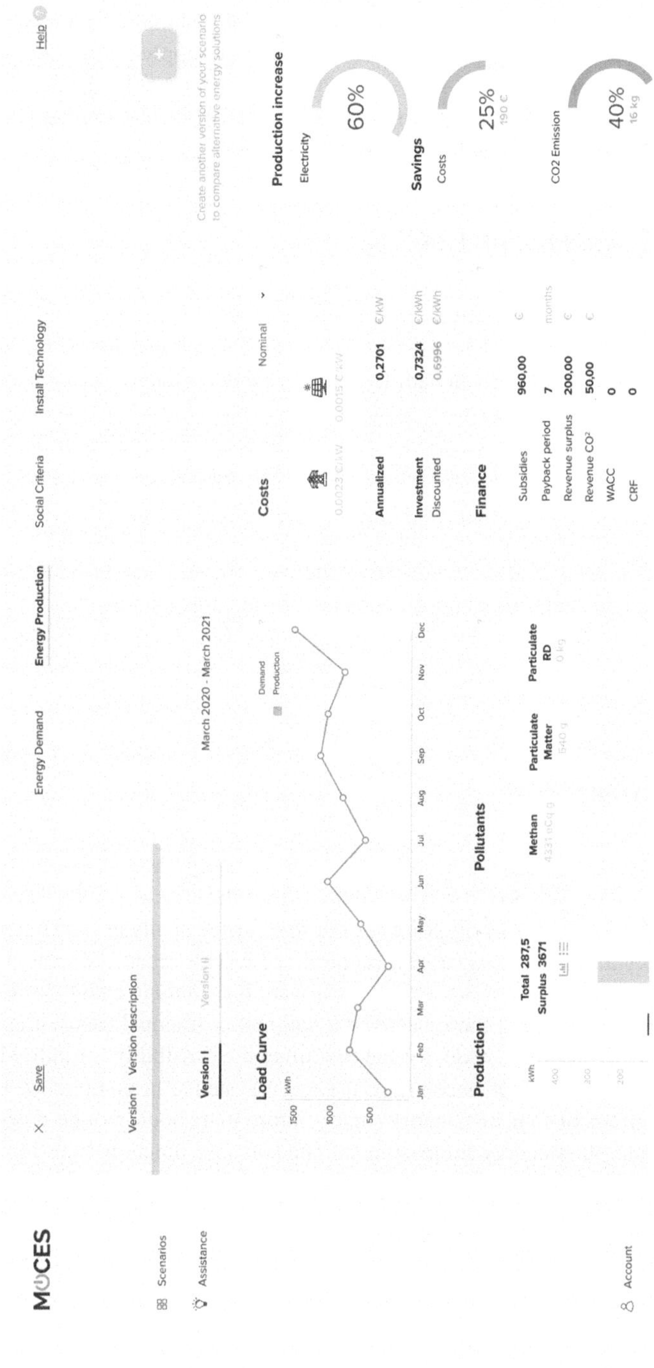

Fig. 7.3 MoCES Results window

clusters (cities, villages, and other networks of energy consumption), (2) to determine the status of electricity access for each cluster, (3) then, to determine, based on the cluster location and population, the optimum electricity supply options.

In Ghana, the Network Planner was developed for planning electricity systems. Network Planner computes costs of different electricity supply options accessible to off-grid communities. Kemausuor et al. (2014) used Network Planner to model access options for off-grid communities of Ghana in a 10-year planning period.

5.2 Value Addition of MoCES

MoCES combines functionalities that position it a step further existent energy planning software. The tool focuses on planning energy systems in the city and sub-urban environments. Sustainability in the energy system goes beyond cost optimization and CO_2 reduction, which metrics are tracked in existing energy planning software. Other criteria such as the technology integration in the environment, comfort of use, and access to a database of professional service providers are relevant in inventing an energy system that is tailored to the local context needs.

MoCES is developed for the sub-Saharan Africa region, where both access to energy and efficiency of use are pressing concerns. The software provides functionalities to design decentralized energy systems that integrate both generation potential (e.g. solar irradiation) and service efficiency potential (e.g. lighting retrofit with LED lamps).

6 Conclusion

MoCES addresses the critical need to democratize energy access and transition to sustainability in the sub-Saharan Africa cities. Since the creation of the Edison lamp in 1879, which sets the starting point for the first power plants in the United States (1883), energy access remains largely a centralized service organized around utilities and community grids. Therefore, existing software for planning energy systems was primarily designed to meet these needs at a national or regional (pool of countries) level. With MoCES, we provide, for the first time, the possibility for individual city agents to plan and monitor decentralized energy systems, in compliance with the citizens commitment to a sustainable energy future that should not be delegated to the utilities and other energy professionals. The software functionalities build from:

(a) Recent developments in information technology in terms of human–computer interactions and visual computing; and.
(b) The less recent concept of modelling systems with agents as an alternative to modelling equilibrium and other optimum solutions, which fail to capture the agents learning and adaptation mechanisms.

Individual energy planning supports an access based on informed decisions. In addition, both universal energy access to sustainable energy and liveable cities are goals of the post-2015 Development agenda. MoCES provides a medium that supports the contribution of citizens to achieving these goals.

References

Barles, S. (2009). Urban metabolism of Paris and its region. *Journal of Industrial Ecology, 13*(6), 898–913. https://doi.org/10.1111/j.1530-9290.2009.00169

Batty, M. (2005). *Cities and complexity: Understanding cities with cellular automata, agent-based models, and fractals*. MIT Press.

Beeck, N. V. (1999). Classification of Energy Models. *FEW Research Memorandum;, 777*. Tilburg, Netherlands: Tilburg University. https://pure.uvt.nl/ws/portalfiles/portal/532108/777.pdf

Benedict, C. (2013). *Case studies on energy subsidy reform: Lessons andimplications*. International Monetary Fund. Retrieved from https://www.imf.org/external/np/pp/eng/2013/012813a.pdf

Bertalanffy, L. v. (1969). *General system theory*. George Braziller.

Bertheau, P., Cader, C., & Blechinger, P. (2016). Electrification modelling for Nigeria. *Energy Procedia, 93*, 108–112. https://doi.org/10.1016/j.egypro.2016.07.157

Focardi, S. M. (2015). Is economics an empirical science? If not, can it become one? *Frontiers in Applied Mathematics and Statistics, 1*. https://doi.org/10.3389/fams.2015.00007.

Georgescu-Roegen, N. (1971). *The entropy law and the economic process*. MA University Press.

Grubler, A., Bai, X., Buettner, T., Dhakal, S., Fisk, D. J., Ichinose, T., ... Weisz, H. (2012). Chapter 18: Urban EnergySystems. In A. Grübler (Ed.), *Global energy assessment–toward a sustainable future* (pp. 1307–1400). Cambridge University Press. Retrieved from https://iiasa.ac.at/web/home/research/Flagship-Projects/Global-Energy-Assessment/Chapte18.en.html

Hayek, F. v. (1948). *Individualism and economic order*. Routledge Press (UK), University of Chicago Press (US).

Kemausuor, F., Adkins, E., Adu-Poku, I., Brew-Hammond, A., & Modi, V. (2014, April). Electrification planning using network planner tool: The case of Ghana. Energy for Sustainable Development, 19, 92–101. https://doi.org/10.1016/j.esd.2013.12.009.

Messner, S., & Strubegger, M. (1995). *Model-based decision support in energy planning*. International Institute for Applied Systems Analysis. Retrieved from http://pure.iiasa.ac.at/id/eprint/4477/1/WP-95-119.pdf

Ringkjøb, H.-K. M., Haugan, P., & Solbrekke, I. M. (2018). A review of modelling tools for energy and electricity systems with large shares of variable renewables. *Renewable and Sustainable Energy Reviews*, 440–459. https://doi.org/10.1016/j.rser.2018.08.002

Schneider, F. (2012). *The shadow economy and work in the shadow: What do we (not) know?* Institute for the Study of Labor. Retrieved from http://ftp.iza.org/dp6423.pdf

Schrattenholzer, L. (1981). *The energy supply model MESSAGE*. International Institute for Applied Systems Analysis. Retrieved from http://pure.iiasa.ac.at/id/eprint/1542/

Shukla, P. R. (1995, August). Greenhouse gas models and abatement costs for developing nations: A critical assessment. Energy Policy, 23(8), 677–687. https://doi.org/10.1016/0301-4215(95)00062-N.

Transform Africa. (2017). *Transform Africa summit: Smart cities*. UK. Retrieved from https://www.uclg.org/en/media/events/transform-africa-summit-smart-cities

Trotter, P., McManus, M. C., & Maconachie, R. (2017, July). Electricity planning and implementation in sub-Saharan Africa: A systematic review. *Renewable and Sustainable Energy Reviews, 74*, 1189–1209. https://doi.org/10.1016/j.rser.2017.03.001

Urban, F., Benders, R., & Moll, H. C. (2007). Modelling energy systems for developing countries. *Energy Policy, 35*(6), 3473–3482. https://doi.org/10.1016/j.enpol.2006.12.025

Chapter 8
Hidden Costs of Decarbonizing Utility Generation: Investment on Grid Stability and Contribution of Renewable Energies

C. A. Mbodji, D. Diouf, L. Ndour, A. L. Fall, and A. Maiga

Abstract How much the energy transition will cost is yet to be determined. In the electricity sector, studies continue to show trends of decreasing costs for various renewable energy technologies such as solar photovoltaic and wind. Recurring recommendations are made from international organizations to phase-out energy subsidies on conventional fuels in order to make competitive electricity generation with renewable energy. In this chapter, the authors demonstrate how electricity generation with renewable resources can contribute to the decarbonization of utility generation. Considering the examples of Ghana and Senegal. They propose a new approach with five pillars to decarbonize electricity supply from the grid in the West Africa region.

Keywords Grid stability · Decarbonization · Renewable energy · Generation costs

1 Introduction

Thermal power plants powered by fossil fuels such as coal, oil, and natural gas continue to be predominant in electricity generation worldwide, which highly affects the sector's carbon footprint. Countries such as the Democratic Republic of Congo, Iceland, Albania, and Norway, which have a relatively high share of their electricity

C. A. Mbodji (✉) · D. Diouf · A. Maiga
Laboratoire Electronique, Informatique, Télécommunications et Energies Renouvelables, Gaston Berger University, Saint-Louis, Senegal
e-mail: mbodji.cheikh-ahmadou@ugb.edu.sn; djicknoum.diouf@ugb.edu.sn; amadou-seidou.maiga@ugb.edu.sn

L. Ndour
Development Specialist, Nairobi, Kenya
e-mail: amadou.ndour@responsAbility.com

A. L. Fall
Global Green Growth Institute, Dakar, Senegal
e-mail: amadou.fall@gggi.org

A. Fall, R. Haas (eds.), *Sustainable Energy Access for Communities*,
https://doi.org/10.1007/978-3-030-68410-5_8

produced with renewable energy resources, exhibit the lowest carbon-electricity generation rates (IRENA, 2018). The decarbonization of electricity generation from utilities is necessary to reduce the environmental strain of energy production, especially in Africa, where the production and demand of electricity increase exponentially.

According to the International Energy Agency, More than half a billion people will be added to Africa's urban population by 2040 (…). These growing urban populations mean rapid growth in energy demand, which makes the projected demand for oil of the continent by 2040 to be higher than that of China and second only to that of India. (IEA, 2019). In 2017, the total installed capacity of renewable energy for electricity generation in Africa was 42 GW. Analysis shows that the continent can meet half of its electricity demand equivalent to a capacity of 310 GW with renewable energy resources by 2030 (IRENA, 2019a, b, c). Therefore, the region urgently needs to improve its network infrastructure while decarbonizing the grid with clean energy technologies. However, the intermittency of renewable energy sources often necessitates the use of a baseload production, which in most cases, is powered by thermal resources. The penetration of intermittent renewable energy in the grid increases its instability and causes dispatching challenges to the utilities. In addition, conventional energy technologies, which are oil, coal, and natural gas, remain attractive (IRENA, 2019a, b, c), especially for developing countries that import renewable energy technology. Another problem relates to the selection of context-specific technologies in grid extension. These problems, among others, raise the question of the opportunity to invest in grid stabilization.

The previous studies on grid decarbonization mainly focus on the contribution of renewable energies to global value chains and the issue of grid stability. Considering sample countries in West Africa, this chapter explores how electricity generation with renewable energy can contribute to decarbonization of the grid while preserving its stability. The chapter is organized as follows: Section 2 reviews a couple of methods proposed to decarbonize electricity generation, as well as pathways to control grid stability. There, we propose a method tailored to the West African context in making generation with renewable energy more competitive. Section 3 presents the results of our analysis of investment costs associated with grid stability, generation with renewable energy, and reduction of the environmental strain. Section 4 discusses how the proposed method could contribute to making electricity generation with renewable energy more competitive in West Africa. The study's objective is to advance knowledge on requirements to decarbonize utility generation and to phase-out sustainably subsidies on conventional electricity generation.

2 Methodological Approach

Several authors studied pathways to decarbonize the electricity generation of utilities. Therefore, several approaches and methodologies are available in the literature. Our study builds from literature review, and proposes a new approach to decarbonize electricity supply from the grid in West Africa.

The Intergovernmental Panel on Climate Change (2015) identified, with a consensus view of 830 scientists, engineers, and economists from 80 countries, four requirements to reach carbon neutrality in electricity generation by utilities. Those are (a) decarbonization of electricity generation; (b) massive electrification (using electricity from renewable sources) and switch to low-carbon fuels; (c) greater efficiency and less waste in all sectors; (d) improved carbon sinks such as forests, vegetation, and soil.

Similarly, the ASPEN Institute (Ballentine & Connaughton, 2019) expressed five basic elements for achieving deep decarbonization of the electricity network:

(a) Use energy efficiency to the maximum degree in order to reduce energy demand
(b) Electrify energy services as much as possible, including heat, transportation, and industrial processes
(c) Use zero-carbon fuels in the areas that cannot be effectively connected to the grid
(d) Use carbon capture, utilization and storage (CCUS) and carbon dioxide removal techniques in areas where fossil fuels are still needed; and
(e) Decarbonize the electricity supply

The World Bank Group (2015) proposed three (3) principles of a decarbonization scheme that can be adapted in each country:

(a) To define a long-term target for the electricity sector that is consistent with decarbonization objectives and to design short-term sector-specific plans that contribute to the long-term target.
(b) To set a policy package that triggers changes in investment patterns and technologies like carbon pricing, which is an efficient way to raise revenues that could support sustainability or help reduce other taxes (e.g. on imports of renewable energy technologies).
(c) To bear in plans the political economy and smooth transition needs with a policy package that is more attractive to a majority of people.

The High-Level Platform for Sustainable Energy Investments in Africa (European Commission, 2019) proposes to reform fossil fuel subsidies specifically, alongside the introduction of decarbonization policies, circular economy practices, and strong environmental and social standards that align with energy efficiency, environmental protection and emissions' performance standards.

Hirth and Steckel (2016) used the power market model EMMA, which is a techno-economic system optimization model (cost minimisation), to evaluate the impact of capital costs and carbon prices on the deployment of renewable energy and other low-carbon technologies in the grid, while accounting for value differences and system costs. They found that high capital costs can significantly reduce the effectiveness of carbon pricing. For instance, if carbon emissions are priced at USD 50 per metric ton and the Weighted Average Cost of Capital (WACC) is 3%, the cost-optimum electricity mix comprises 40% renewable energy.

After having explored the above-listed principles and methods, we propose the following package with five pillars as a contribution to reach decarbonization of electricity supply from the grid in the West Africa context.

Firstly, countries should set policies that promote renewable energy with practical measures to reallocate subsidies from fossil fuels to renewable energy technologies.

Secondly, decentralized mini-grids powered with locally available renewable energy resources should be prioritized in electrification programmes and in grid extension schemes.

Thirdly, countries should invest in computer based-modelling tools to model and anticipate grid instability, notably large disturbances that are due to renewable energy resources intermittency and short circuits of high voltage equipment.

Fourthly, each economic sector should be required to have specific plans of electricity supply with zero-carbon fuels, including renewable energy resources such as waste recycling in industry processes and solar rooftops in commercial buildings. The sectoral plans should also include energy efficiency measures to alleviate the demand.

Fifthly, an increase of the carbon sinks through afforestation and greening of urban areas should complement efforts in the electricity network.

3 Discussion of Findings

The International Renewable Energy Agency (2019) states that the share of renewable energy in the electricity supply could increase to 50% by 2030 and to 73% by 2050 in Africa. However, the investment requirements to stabilize the grid continue to hinder the process. Indeed, underinvestment in the electricity transmission and distribution networks leads to serious inefficiencies, with electricity losses averaging 14.8% of the production (Bose, 2003). For a long time, the investment cost of renewable energy technologies has been the major obstacle to the decarbonization of electricity generation by utilities. Nowadays grid stability is at top priority because the investment and production costs of renewable energy technologies are constantly decreasing, while network infrastructures remain underequipped. Table 8.1 provides an overview of the cost of electricity generated from different renewable energy technologies and statistical variations. IRENA states that onshore wind and solar PV are set by 2020 to consistently offer a less expensive source of new electricity than the least-cost fossil fuel alternative, without financial subsidies.

Addressing the issue of grid stability requires more human than financial resources. Context-specific solutions should be privileged, taking into account the available conventional and renewable energy resources. Considering the grid

Table 8.1 Global electricity costs in 2018 (IRENA, 2019a, b, c)

	Global weighted average cost of electricity (USD/kWh) 2018	Cost of electricity: fifth and 95th percentiles (USD/kWh) 2018	Change in cost of electricity 2017–2018
Bioenergy	0.062	0.048–0.243	−14%
Geothermal	0.072	0.060–0.143	−1%
Hydropower	0.047	0.030–0.136	−11%
Solar photovoltaic	0.085	0.058–0.219	−13%
Concentrating solar power	0.185	0.109–0.272	−26%
Offshore wind	0.127	0.102–0.198	−1%
Onshore wind	0.056	0.044–0.100	−13%

Table 8.2 Senegal's Electricity Production Mix in 2017 (SENELEC, 2017)

Energy Source	Installed Power	Share (in %)
Thermal (heavy fuel & diesel)	782.83 MW	74.78%
Thermal (steam generators)	87.5 MW	8.36%
Solar photovoltaic (PV)	101.5 MW	9.70%
Hydropower	75 MW	7.16%
Total installed	1046.83 MW	2139,925 GWh

characteristics of Senegal and Ghana, we test how our approach could contribute to decarbonization of the electricity generated by the utilities in these countries.

In Senegal, thermal sources (diesel and heavy fuel) are still predominant in electricity generation. Table 8.2 provides an overview of the electricity production figures in 2017 from the utility (SENELEC).

In 2017, renewable energy in the electricity generation mix of the utility represented 16.9%. In 2018, the total installed capacity of solar photovoltaic and hydropower represented a renewable energy penetration equivalent to 21% of the total installed capacity, with 17% (157 MWp) from solar PV (CRSE, 2019). The grid emission factor in Senegal was 0.56 tCO_2/MWh in 2018 (UNFCCC, 2017). In comparison, the average grid emission factor in Europe was 0.29 tCO_2/MWh and 0.08 tCO_2/MWh for Austria (European Environment Agency, 2018). This observation confirms the urgency to decarbonize electricity generation in Senegal. This decarbonization process can be incentivised by applying the first measure of our 5-pillar package, namely by shifting subsidies from existent and future fossil fuels-powered generation to promote investment in locally available renewable energy resources. In Senegal, these transfers should primarily target solar and wind resources that are available in many parts of the country. It is proven that onshore wind and solar PV are now less expensive sources for electricity generation than some fossil fuels, including fuel oil (IRENA, 2019a, b, c).

Table 8.3 Ghana's Electricity indicators in 2017 (Energy Commission of Ghana, 2018)

Energy Indicator	Unit	Value
Total primary energy supply	Ktoe	9614
Electricity generation form the grid	GWh	14,068
Electricity from the grid consumed	GWh	12,091
Grid emission factor	Tonnes CO_2/MWh	0.43

The integration of solar PV systems into the grid took place mainly between early 2016 and 2019. During that period, the solar PV capacity connected to the grid increased from 2 MWp to 157 MWp with the commissioning of eight (8) solar photovoltaic power plants: Bokhol (20 MWp), Malicounda (22 MWp), Santhiou Mékhé (30 MWp), Tenmérina (30 MWp), Kahone (20 MWp), Sakal (20 MWp), Diass (15 MWp) (ASER, 2019). This growth is driven by a political commitment to materialize provisions of the 2001 Law that promotes the diversification of the electricity generation mix by using locally available renewable energy resources. The external factor that supported the political action was the constant decrease in international markets of renewable energy technologies. The Senegal example confirms two levers are critical in grid decarbonization with renewable energy technologies, namely the combination of an internal factor, i.e. political will and an external factor, i.e. reduction of technology costs.

The continuing decrease of solar PV and wind energy technologies is an incentive for Senegal to continue investments in these technologies. Still, stability of the grid requires to integrate baseload generation technologies, and since 2018, the government has started investing in coal-powered power plants. This requirement could be balanced by another pillar of our approach, which is the development of carbon sinks. Afforestation initiatives to create carbon sinks should target urban and peri-urban areas to compensate emissions from the coal power plant and to target a net-zero network that combines the power plants and carbon sinks. Different vegetal species can be considered for carbon sink (e.g beefwood), and the areas of Bargny and Sebikotane located near the coal power plants should be prioritized for reforestation. Our second pillar, i.e. using computer-assisted modelling tools to control grid stability, can be considered in order to integrate decentralized energy generation systems in the fisheries and agriculture transformation industries installed in the vicinities of the coal power plants. Indeed, urban gardening and fish refrigeration in the coastal area of Senegal use decentralized energy generation facilities for water pumping and electrification. The integration of these systems into the network requires control tools that are flexible enough to adapt context-specific parameters such as power and operation hours.

Hydropower installations on the Volta River are predominant in the Ghana's electricity supply mix. The Ghana generation capacity was 2450 MW in 2015, with 54% representing hydropower systems and 46% representing thermal generation systems (USAID, 2015). Table 8.3 provides an overview of figures related to the Ghana electricity generation mix.

The predominance of hydropower in the electricity supply mix explains the relatively low grid emission factor, compared to Senegal. Considering the geographic position of Ghana, an investment in carbon sinks through initiatives such as REDD+ could also be considered to compensate for emissions from thermal generation power plants.

The investment in additional renewable energy resources could target rural areas of the country with off-grid energy solutions such as decentralized grids in the replacement of grid extension schemes. Another pillar of our approach for decarbonizing the grid in Ghana should target the mining industry, especially small-scale mining businesses, which could significantly alleviate the grid.

From these two examples of countries having different energy mixes and grid emission factors, we see applicability of our approach, at diverse degrees, as a contribution to decarbonize electricity generation by the utility.

The cost of decarbonization could be further reduced in both countries with ambitious policies that provide local governments with competencies to manage energy systems established in their territories, such as third-party access to the grid. The implementation of these policies should be sustained by the development of tools that monitor deficit or excess of production due to intermittency of renewable energy resources and the investment in baseload systems using renewable energy resources, which can be biomass or hydropower. The design of a consistent agenda, including the proposed five-pillar package to decarbonize the grid, will further support local governments' contributions to current initiatives for transition to energy sustainability in rural and urban communities of the West Africa region.

4 Conclusion

The West Africa region is endowed with renewable energy resources, including solar photovoltaic and hydropower, which national utilities are increasingly using to generate electricity and improve access through the interconnected grid. In addition, mass electrification using decentralized mini-grids in West Africa proved that renewable energy technologies are applicable solutions to improve access. For instance, the rural electrification programme (ERIL) in Senegal, which promotes decentralized solar photovoltaic mini-grids in Senegal's countryside, rather than the extension of the interconnected grid, increased the rate of electrification in rural areas to 42% in 2018 compared to 16% in 2007 (ASER, 2019). However, the challenge remains on how to stabilize the grid while transitioning to low-carbon electricity generation, whether it is in urban or rural communities. The five-pillar approach proposed in this chapter is a contribution to a holistic solution that decarbonizes the grid while preserving its stability. The contribution of local government to the process requires that the pillars be adapted to communities' context; baseload generation to stabilize the grid can use either biomass, including recycled waste-to-energy, or hydropower, depending on the context. This adaption should mobilize local and national competencies, as well as the research community, in innovating with adaptable tools that simulate and monitor decarbonization schemes in order to meet the energy access and transition to sustainability objectives of countries in West Africa region.

References

ASER. (2019). *Elaborationd'un plan opérationnel pour la mise en œuvre d'un programme d'électrificationrurale*. Agence Senegalaise d'Electrification Rurale.

Ballentine, R., & Connaughton, J. (2019). *Decrabozing the electricity sector & beyond*. Washington D.C. Retrieved from https://assets.aspeninstitute.org/content/uploads/2019/07/2019-Energy-REPORT.2.pdf

Bose, A. (2003). Power system stability: New opportunities for control. In D. Liu & P. J. Antsaklis (Eds.), *Stability and control of dynamical systems with applications. Control engineering* (pp. 315–330). Springer. https://doi.org/10.1007/978-1-4612-0037-6_16

Commission de Regulation du Secteur de l'Energie. (2019). *Bilan des activites de SENELEC*. CRSE.

Energy Commission of Ghana. (2018). *National Energy Statistics 2008–2017*. Energy Commission of Ghana.

European Commission. (2019). *Scaling-up energy Investments in Africa for inclusive and sustainable growth*. Africa–Europe High–Level Platform for Sustainable Energy Investments in Africa. Retrieved October 2020, from https://ec.europa.eu/international-partnerships/system/files/report-africa-europe-high-level-platform-sei_en.pdf

European Environment Agency. (2018). *CO_2 Emission Intensity*. Retrieved September 2020, from eea.europa.eu/data: https://www.eea.europa.eu/data-and-maps/daviz/co2-emission-intensity-1#tab-chart_1_filters=%7B%22rowFilters%22%3A%7B%7D%3B%22columnFilters%22%3A%7B%22pre_config_ugeo%22%3A%5B%22European%20Union%20(28%20countries)%22%5D%7D%7D

Hirth, L., & Steckel, J. C. (2016). The role of capital costs in decarbonizing the electricity sector. *Environmental Research Letters, 11*. https://doi.org/10.1088/1748-9326/11/11/114010

IEA. (2019). *Africa energy outlook 2019*. IEA. Retrieved from https://www.iea.org/reports/africa-energy-outlook-2019

Intergovernmental Panel on Climate Change. (2015). *AR5 synthesis report: Climate change 2014*. IPCC. Retrieved from https://archive.ipcc.ch/report/ar5/syr/

IRENA. (2018). *RenewableElectricity capacity and generation statistics*. IRENA.

IRENA. (2019a). *RenewablePower generation costs in 2018*. International Renewable Energy Agency.

IRENA. (2019b). *Scaling up renewable energy deployment in Africa: Impact of IRENA's engagement*. IRENA. Retrieved from https://www.irena.org/-/media/Files/IRENA/Agency/RegionalGroup/Africa/IRENA_Africa_impact_2019.pdf?la=en&hash=EECD0F6E8195698842965E63841284997097D9AA

IRENA. (2019c). *World energy outlook 2019*. IEA. Retrieved from https://www.iea.org/reports/world-energy-outlook-2019/oil#abstract

SENELEC. (2017). Statistiques de la production d'electricite. Récupéré sur Senelec.sn: http://www.senelec.sn/production/

UNFCCC. (2017). *Standardized baseline: Grid emission factor for west African power Pool.* United Nations framework convention on climate change. Retrieved September 2020, from https://cdm.unfccc.int/filestorage/e/x/t/extfile-20170228152153668-ASB0034_PSB0033.pdf/ASB0034_PSB0033.pdf?t=QkZ8cWtraHptfDAPzCcpK2e5eVw9_IG9XNbA

USAID. (2015). *What power Africa means for Ghana.* U.S. Agency for International Development.

World Bank Group. (2015). *DecarbonizingDevelopment, three steps to a zero-carbon future.* Washington D.C.

Chapter 9
Modelling Sustainable Energy Transition for Cities: Case Studies of LEAP, ENPEP-BALANCE, and MoCES

A. Fall and C. A. Mbodji

Abstract The city of Dakar, in Senegal, is endowed with solar photovoltaic, waste-to-energy, and wind energy resources that could support the transition to energy sustainability in the production of electricity for the interconnected grid. Ordinarily, energy planners use modelling software to simulate the integration of this potential into the electricity generation network. In 2018, Ringkjøb et al. identified 75 energy modelling software. The study simulates the integration of the city renewable energy potential in the electricity generation network, using three energy planning software. The main finding is the approach, which is embedded in the software programming interface, affects the model results in terms of competitiveness of renewable energy technologies when compared to existing energy technologies.

Keywords Energy planning · Energy software · System simulation · Computational equilibrium · Agent-based modelling

1 Introduction

Since the 1970s, energy software has been used to describe, plan, and monitor energy systems. The first energy modelling software appeared between the oil crisis of 1973 and the energy crisis of 1979, with the Market Allocation model known as MARKAL (1978) and the Wien Automatic System Planning Package known as WASP III (1979). Since then, the number of energy software has constantly increased, and in 2018, Ringkjøb et al. (2018) identified 75 energy modelling

A. Fall (✉)
Energy Economics Group TU Wien, Vienna, Austria
e-mail: fall@eeg.tuwien.ac.at

C. A. Mbodji
Laboratoire Electronique, Informatique, Télécommunications et Energies Renouvelables, Gaston Berger University, Saint-Louis, Senegal

A. Fall, R. Haas (eds.), *Sustainable Energy Access for Communities*,
https://doi.org/10.1007/978-3-030-68410-5_9

software. However, the weight of the model's approach on the energy systems computed is still unclear.

A previous study (Fall, 2020) assessed the potential of renewable energy resources in Dakar and demonstrated a technically achievable potential on solar photovoltaic, waste-to-energy, and wind energy resources. This study uses the figures associated with this potential to explore how different modelling approaches affect energy metrics depending on the software used. The study hypothesized that the modelling approach affects the city energy metrics, and different software can result in different portraits of the future energy system, which, together, affect policy recommendations for transition to sustainability. Most publications on energy modelling software emphasize the modeller's responsibility as a probable cause for this discrepancy; the selection of a modelling software only depending on the user's preferences. However, these preferences do not necessarily translate into an informed decision considering all alternatives, but more than often relates to the accessibility of the modelling software (free download or paid licence), technical knowledge of model operations (inputs and outputs), previous exposure to the software environment (e.g training), etc.

The first objective of the chapter is to introduce MoCES, which approach considers individual agents having rules in planning city energy systems. With this software, the authors bring innovation in the field of energy modelling as agent-based modelling is not frequently used in planning energy systems, despite its popularity in economics. The second objective of the paper is to compare MoCES computation results with two renowned energy planning software that feature different approaches: LEAP and ENPEP-BALANCE. The rationale of this study is to quantify the tractable implications of the modelling approach in planning the transition to energy sustainability for the city.

2 Methodological Approach

The study compares three sample energy software in modelling a future energy system from a single reference energy scenario (2016) and using the same renewable energy technologies. The sample city is Dakar, and figures associated with the city potential on renewable energy capacities (Fall, 2020) are:

- Solar photovoltaic (39.6 MW)
- Waste-to-energy (689.95 GW)
- Wind energy (5951.4 MW)

The sample software in our study has different approaches in modelling future energy systems, which are system simulation for LEAP, computational equilibrium for ENPEP-BALANCE, and agent-based modelling for MoCES. MoCES is an energy planning software designed within the framework of this study.

2.1 *Long-Range Energy Alternatives Planning System: LEAP (Stockholm Environment Institute, 2020)*

LEAP's approach is a simulation of energy scenarios from a baseline (Current Accounts) to the end of the scenario period (End-Year). The user inputs data of the baseline year in the Current Accounts and includes changes in scenarios in the form of numerical values. In the analysis window (View) of LEAP, the user can input the system energy demand, transformation (conversion from primary energy resources to secondary energy production), and resources potential. The software features a Technology and Environmental Database (TED) that enables the user to connect appliances, devices, fuels, and technologies entered in the model to a database that includes metadata such as pollutant emissions of these appliances, devices, fuels, and technologies. LEAP also provides an option to optimize the energy system modelled.

2.2 *Energy and Power Evaluation Programme: ENPEP-BALANCE (Argonne National Laboratory, 2019)*

ENPEP-BALANCE's approach is the optimization of an energy network made of energy production (resources), conversion, transport, distribution, and end-use nodes, as well as the flows of energy and fuels among those nodes. Opposite to LEAP and MoCES, the software interface is an empty workspace that the modeller fills in configuring an energy network with embedded nodes and links. ENPEP-BALANCE simultaneously finds the intersection of supply and demand curves for all energy supply sources and all energy uses included in the network. Equilibrium is reached when the model finds a set of market-clearing prices and quantities that satisfy all relevant equations and constraints entered by the user.

2.3 *Modelling Cities Energy Systems: MoCES (Fall et al., 2020)*

MoCES is an agent-based modelling software that aims at computing the city energy network based on individual agents' energy choices. At city and district levels, MoCES computes the total energy end-use considering the average energy per capita and the population, which is similar to excel-based calculations. The model trades off its basic features in these levels of analysis with a more detailed interface for individual agents, whether it is a residence, commerce, or industrial building.

- On the demand side, MoCES computes the demand of the building energy services, including lighting, cooling, heating, and cooking, while considering these variables: number of plugging appliances and devices in the building, wattage (W), and usage time per day (hours).
- On the supply side, MoCES computes the energy of each production technology accessible in the building and available in the catalogue that includes solar photovoltaic, solar thermal, wind energy, and waste recycling to energy while considering these variables: power (kW), efficiency, capacity factor and operational hours per period (hours). The user can also compute energy from the grid in this window with metering data for benchmark purposes.

Figure 9.1 depicts the MoCES' demand window.

2.4 Data Sources

The data in the reference energy scenario of the model (2016) are from three material sources:

(a) Report of the Senegal Information Energy System (SIE 2016) that has been collected from the Ministry in charge of Energy (Ministere en charge des Energies, 2018).
(b) Utility report on the electricity sector in 2016 (SENELEC, 2017) that is accessible online.
(c) The matrix of data collected during the survey on citizens' energy behaviour performed in different districts of Dakar between November 2018 and May 2019, which was conducted within the framework of the project Sustainable Energy Access for Sustainable Cities (SEA4cities). Annex 1 provides supplementary materials on the survey.

3 Data and Results

From a single reference energy scenario (Scenario 1), which is a simplified—modelled—version of the city energy network in 2016, we model two future energy scenarios (2017–2030) using the sample modelling software:

- Integration of the renewable energy potential in the supply mix (scenario 2)
- Demand-side-management in the residential sector (scenario 3)

Fig. 9.1 MoCES Demand Window

3.1 Reference Energy Scenario (RES)

Figure 9.2 displays the simplified city energy network in the baseline year (2016).

Energy in our network flows from primary energy resources (bottom) to end-use in the residential, commercial and industry sectors (top). We provide detailed information on the network input parameters in the following paragraphs.

3.1.1 Primary Energy Resources

Table 9.1 displays the quantities, conversion factors, and costs in EUR per tonne oil equivalent (toe) of primary energy resources in the network.

The conversion rate USD to EUR is 0.9, which was the average rate in 2016 (Statista, 2020).

Considering reforestation cost of EUR 991 per ha and a charcoal land intensity of 8333.3 kg per ha (Ministere en charge des Energies, 2018), the energy content of the wood species in the study of 20.9 MJ per kg (Fall, 2020), and a wood to charcoal conversion efficiency of 20%, the wood price is 23.8 per metric ton.

Among the primary energy resources in the network, crude oil and coke are imported. Natural gas and wood are domestic resources.

Oil Refining

From imported crude oil (1.1. million tonnes), the local refineries produce diesel, fuel oil, and LPG with an efficiency rate of 88%. The model assumes that any shortfall of diesel, fuel oil and LPG demand in the network is met through imports.

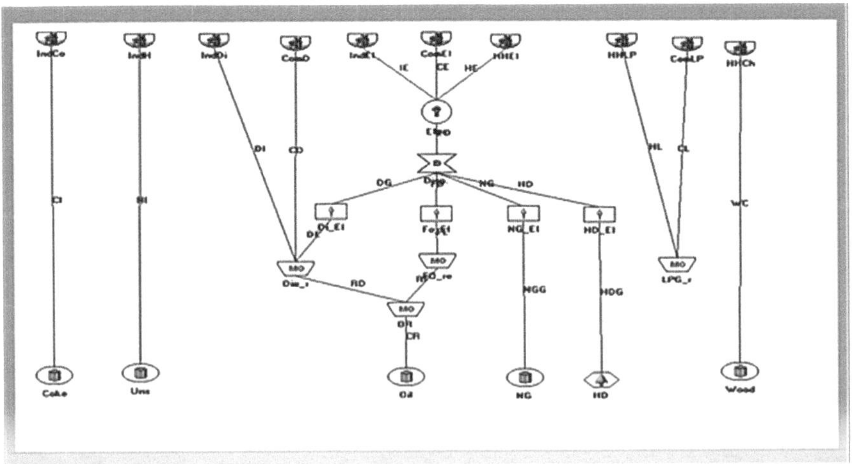

Fig. 9.2 RES City energy network (created with ENPEP)

Table 9.1 RES Primary Energy Resources (2016)

Resources	Consumption (toe)[a]	Price (USD)[b]	Conversion factor (toe)	Price (USD/toe)	Price (EUR/toe)[c]
Coke	416,566	120.59 per tonne	1 tonne = 0.64 toe	187.5	168.75
Crude oil	1,102,202	43.74 per barrel	1 barrel = 0.14 toe	312.43	281.19
Natural gas	416,604	2.87 per million BTU	1 million BTU = 0.026	108.8	97.92
Wood	1691.86	23.8 per tonne	1 tonne = 488 (10–3)	49.79	44.81

[a]Figures in the column are from SIE 2016 (Ministere en charge des Energies, 2018)
[b]Fuel prices are the average of 2016 prices in the international market of commodities downloaded from Statista on February 2020. For wood, we considered the reforestation cost of EUR 991 per ha (Ministere en charge des Energies, 2018)
[c]This column uses information in precedent columns to calculate the unit cost of each primary energy resource in the system per tonne oil equivalent

Table 9.2 Charcoal production parameters

	Km^2	ha
Green areas in the city	161.9	16,190
Green areas with beefwood [a]		81
Charcoal intensity (ha/kg)[b]		0.12
Charcoal potential (kg)		674,583.33
Equivalent wood potential (kg)		3,372,916.67
Equivalent wood energy (MJ)		70,493,958.3
Conversion factor (toe/MJ)		41,868
Equivalent wood production (toe)		1691.86

[a] The study assumes beefwood as wood species for the charcoal production. We were not able to find in the literature a specific figure on beefwood share in the woodland of the city. Therefore, we assume 0.5% share (over 200 species available in the woodland). The share could also be understood as a sustainable wood harvesting, where 0.5% of the woodland should be harvested every year
[b] The charcoal intensity in kg per ha is from SIE 2016 (Ministere en charge des Energies, 2018)

Charcoal Production

In our simplified energy network, we consider charcoal, because it can be replaced by other fuels in the network, i.e. electricity and LPG in providing the same cooking service. Charcoal comes from the local production of wood that uses kiln with an average efficiency of 20%. Table 9.2 presents information on the wood to charcoal conversion scheme in our model. The model assumes that any shortfall of charcoal demand is met through imports.

3.1.2 Energy Conversion

Electricity Transmission and Distribution
According to the utility (CRSE, 2017), the losses in the electricity transmission and distribution in the interconnected grid were in 2016:

- Transport: 1.2%
- Distribution: 16%

Energy Generation Table 9.3 provides parameters related to secondary energy production (conversion from primary resources) in thousand tonnes oil equivalent (ktoe).

We assume 62% as the share of the city in the interconnected grid generation. This figure is the share of electricity from the interconnected grid consumed in the city in 2016 (ANSD, 2019).

The hydropower resource is outside the city, which geographic location (coastal region at about 20 metres above sea level) does not allow the installation of hydropower systems.

With respect to capacities, the study assumes that the overall installed capacity in the interconnected grid is available to supply the city. Then, we enter in the model

Table 9.3 RES Secondary Energy Production in ktoe

Conversion	Production	Availability rate	Efficiency rate	Cost (EUR/kWh)[a]	Cost (EUR/toe)[d]
Diesel (electricity)	144.3	0.88	0.39	0.07	766.86
Fuel oil (steam)	14.2	0.6	0.39	0.09	1053.85
Natural gas	0.33	0.92	0.39	0.17	1922.81
Hydropower	19.2	0.8	0.8	0.032	371.02
Charcoal	46.7		0.2		123.25
LPG [b]	423.3				65.34
Diesel (end-use) [c]	119.8				494.4

[a]The columns production, availability rate, efficiency rate, and costs of conversion of power plants are from the utility (SENELEC, 2017)
[b]The figures for charcoal and LPG are from SIE 2016. LPG is used in 86% of households at a cost of EUR 0.76 per kg (EUR 4.6 for the 6 kg bottle). The energy intensity of LPG is 0.07 toe per household. Charcoal is used in 14% of households. The energy intensity of charcoal is 0.11 toe per household. Charcoal expense is on average EUR 9.3 per household (Ministere en charge des Energies, 2018)
[c]Diesel end-use is a computed figure that reconciles figures of energy end-use in the commercial and industry sectors from SIE (Ministere en charge des Energies, 2018) and our assumptions on the city share of energy consumption from these sectors in the RES. For instance the diesel end-use of commerce is the city commerce end-use estimated as 39.5% of national figures minus other fuels end-use in the sector provided in the SIE report (Ministere en charge des Energies, 2018)
[d]This column converts the unit costs provided in the above-named documents, considering the conversion rate 1 toe = 11,630 kWh

the availability rates provided by the utility for these installations in 2016 and the peak load data per month (Annex 2). The efficiency of thermal units was on average 39% in 2016. The study assumes 80% efficiency for hydropower units.

The dispatching of power plants considers the running cost rule, meaning plants with the lowest generation costs (diesel, fuel oil and steam) are baseload plants, and plants with the highest generation cost (natural gas) are peak load plants.

3.1.3 Energy Demand

Residence

Figure 9.2 shows the energy end-use of households are electricity, charcoal, and LPG. Electricity is used for cooling, lighting, refrigeration, and for the operation of other plugging appliances. Table 9.4 displays the energy intensity of households' electricity services.

For each service, we computed an average of the energy intensity using data from the surveys on Dakar's energy behavior completed in May 2019 (low standard district) and November 2019 (high standard districts). For additional information on the survey, see Annex 1. According to the national statistics agency (ANSD), quoting the utility, the average electricity consumption per household was 1103.9 kWh (ANSD, 2019). Higher figures from the survey data can be explained by the periods of survey, and/or the method used to extrapolate daily consumption averages. For lighting, the average energy intensity per household per year is 409.5 kWh (low standard district) and 602.7 kWh (high standard district). Then the study considers that the city's average is 506.1 kWh per household per year. We use the same method to estimate the energy intensity of other services. The total electricity demand of services in our model amounts to 3066.3 kWh per household per year in the city. Residential grid users in the city were 424,939 in 2016 (CRSE, 2016).

Industry

Total energy consumption of the industry sector was 723 ktoe in 2016 (Ministere en charge des Energies, 2018). Ninety-one (91) per cent of these industries were located in Dakar, according to the 2016 General Survey of Enterprises (ANSD, 2017). Therefore, the energy intensity of industries in our sample city is estimated at 657.9 ktoe (equivalent to 7651.4 GWh), of which 1716.7 GWh of electricity. Other industry energy uses are coke (cement industry), diesel for backup generation and unavoidable steam that results from some industrial (e.g. phosphates) business-as-usual activities.

Table 9.4 Intensity of households' electricity services (in kWh)

Service	Lighting	Cooling	Refrigeration	TVs	Others
	506.1	1153.5	1297.5	73.2	36
Total	3066.3				

Commerce
Among the non-industrial enterprises identified in the 2016 General Survey of Enterprises (ANSD, 2017), 39.5% were located in Dakar. The energy consumption of the commercial sector was 1122 ktoe at the national level (Ministere en charge des Energies, 2018). Therefore, the energy intensity of enterprises in the city is estimated at 443.2 ktoe. Other energy uses of the commerce sector are diesel that fuels backup generators, and LPG for some businesses (e.g. restaurants).

3.2 Renewables in Electricity Generation (Scenario 2)

In scenario 2, we assume:

(a) City population grows at the rate of 3.7% per year; figure provided by the World Bank for Dakar during the period 1990–2018 (World Bank Group, 2019).
(b) Number of households connected to the grid increases by 4.4%, which is an average of the period 2009–2016 computed from the annual utility reports of the period.
(c) The city's renewable energy potential (Fall, 2020) is integrated into the electricity generation mix.

Table 9.5 summarizes figures related to the city's renewable energy potential (Fall, 2020).

Bioenergy potential of the city for electricity generation is made of waste-to-energy. Technologies for conversion of waste to electricity are anaerobic digestion (AD) and dendro liquid energy (DLE).

3.3 Demand-Side-Management in the Residence Sector (Scenario 3)

In scenario 3, the study assumes from the Scenario 2 as baseline, improvements in households' energy behaviour compared to observations during the survey:

Table 9.5 Renewables' potential capacity and costs

Technology	Bioenergy		Solar PV	Onshore Wind
	DLE	AD		
Capacity (MW)	33.6	6	689,945.3	5951.4
Efficiency factor (%)	80	66	21	40
Capacity factor (%)	84	84	17	31
Installation cost (EUR/kW)	517	1890	1609	1609
O&M (EUR/kWh)	0.01	0	0.027	0.01
CO_2 cost (EUR/kWh)	0	0.01		

(a) Lighting energy intensity decreases by 2/3, equivalent to a retrofit of bulbs from halogen (18 Watt) to compact fluorescent light (CFL) (6 W) standard or from CFL to LED (2 W) standard at the constant brightness of 200 lumens per m2.
(b) Cooling energy intensity decreases by 20% equivalent to an increase of the balance temperature point (comfort temperature) by 1degree Celsius. For example a building with constant environmental factors such as air exchange factor, specific heat capacity, and air density will require 164 Wh energy to reach 22 degree Celsius balance temperature point when the outside temperature is 25 degree Celsius. When we increase the balance temperature point to 23 degree Celsius, the cooling energy requirement becomes 131 Wh, meaning a decrease by 20%.
(c) Refrigeration energy intensity decreases by 32% equivalent to retrofit from low standard fridges of 220 Watt to medium standard fridges of 150 Watt.
(d) Energy intensity of TVs, phone, and other appliances decreases by 10% due to manufacture improvement in battery autonomy or sleep mode consumption, without additional action from the user.
(e) All households in the city use LPG for cooking.

4 Discussion of Results

4.1 Reference Energy Scenario (RES)

Figure 9.3 displays the city Reference Energy Scenario (2016) generated with LEAP.

Legend: The energy balance is presented in tonnes oil equivalent, with all input parameters converted with the embedded LEAP units' converter. It is possible to convert the figures to other energy units such as GWh for electricity, using the international Energy Agency online unit converter accessible at https://www.iea.org/reports/unit-converter-and-glossary.

The overall electricity generation in the network was 697.5 ktoe in 2016. This generation mainly relies on diesel (85.8%), and fuel oil (8.5%) produced from crude oil imported by refineries. Hydropower (5.5%) and natural gas (0.2%) complete the electricity generation resources. Energy losses (617.4 ktoe) include refineries, power plants, and network losses. It represents more than the electricity distributed to end-users due to the relatively low efficiency of thermal generation units (diesel, fuel oil and natural gas). As electricity demand for end-use sectors (280.1 ktoe) is more than the system's supply capacities (229 ktoe) after transmission and distribution losses, the network imports 51.1 ktoe electricity to meet the demand of end-use sectors. The heat use of industry comes from domestic resources (natural gas), and there is no import of heat in the RES, despite the fact LEAP displays it with 0 as value.

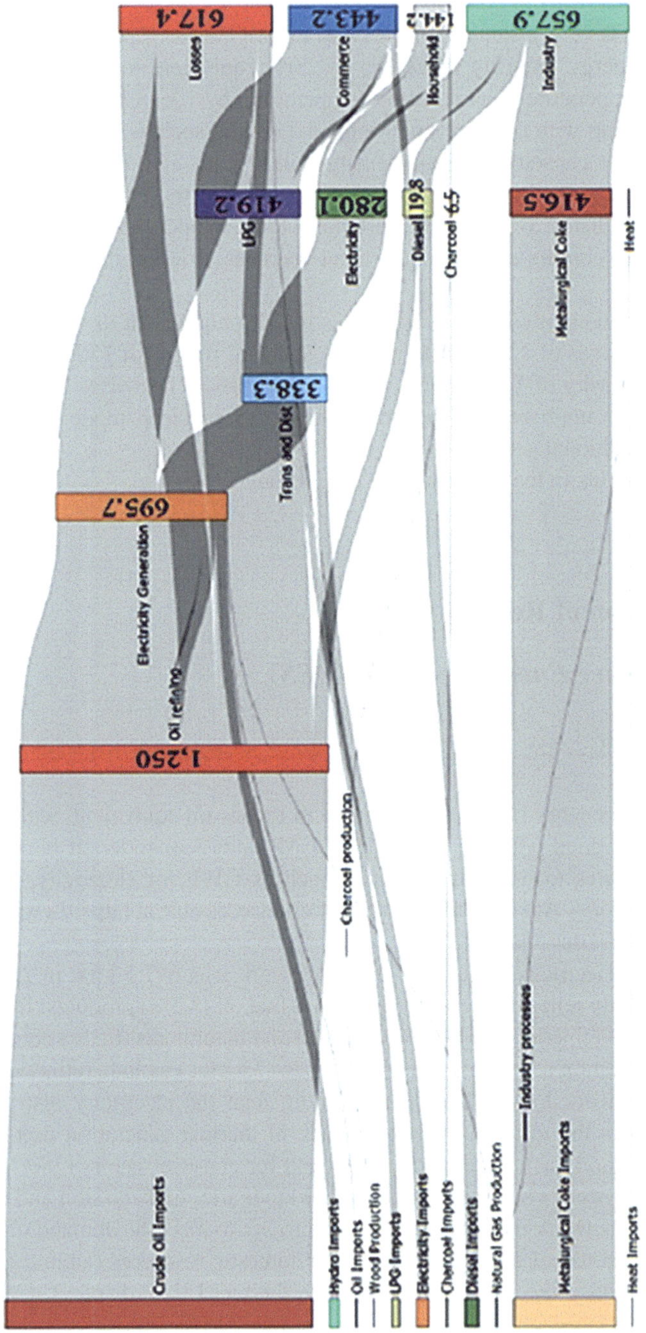

Fig. 9.3 RES-LEAP (in thousand tonnes of oil equivalent-ktoe)

4.2 Renewables in Electricity Generation (Scenario 2)

4.2.1 Leap

Figure 9.4 displays the city energy network under scenario 2 in LEAP.

Electricity generation increased by 179% and supplies the overall city electricity demand (imports = 0). Wind energy resources converted to electricity at the average cost of EUR 8.3 cents per kWh supplies 98.8% of the city electricity demand. This generation cost is higher than the diesel unit cost (EUR 7 cents per kWh) and hydropower unit cost (EUR 3 cents per kWh). However, both technologies run with imported fuel, and LEAP computes results, prioritizing domestic resources per default. In this scenario, crude oil imports decreased by 52.1%, and its refining only produces LPG and diesel for the commerce and industry sectors end-uses.

The remaining electricity generation (1.2%) is from waste-to-energy (anaerobic digestion and dendro liquid energy), which is the cheapest domestic resource. The System cost is EUR 5864.6 mio.

4.2.2 ENPEP-Balance

Figure 9.5 displays the city energy network under scenario 2 with ENPEP-BALANCE.

Legend: The model constant parameters are:

- Premium multiplier that we assumed at 1 to indicate neutrality over fuels available as options. Premium multipliers reflect the preference for a fuel over others. A multiplier greater than 1 raises the price of competing energy products in the market share equation. A multiplier lower than 1 lowers the price of competing energy products in the market share equation.
- Cost sensitivity factor that we assumed at 0.5. The 0 value is an extreme case and indicates the least degree of the fuel share sensitivity to prices. A value above 1 indicates a higher degree of the fuel share sensitivity to relative prices.
- Lag factor that we assumed at 1. The lag value ranges between 0 and 1. A value of 1 indicates there is no lag, and shares respond to current prices. A value of 0 indicates no response to prices, meaning base-year shares are maintained throughout the study period.

In ENPEP, the share of a fuel in the city supply mix is inversely proportional to its cost. The market share of the fuel is its relative price over the sum of all fuels' relative prices. Therefore, the ENPEP model returns a situation where all available fuels are in the supply mix; the cheapest option having the higher share. In 2017, DLE was the cheapest option (EUR 1.9 cents per kWh), but it has a capacity constraint (33.6 MW) that limits generation to 17.1 thousand tonnes oil equivalent per year. The other waste-to-energy technology (AD) was also used at full capacity (6 MW) as the third-cheapest generation option. In between, hydropower import

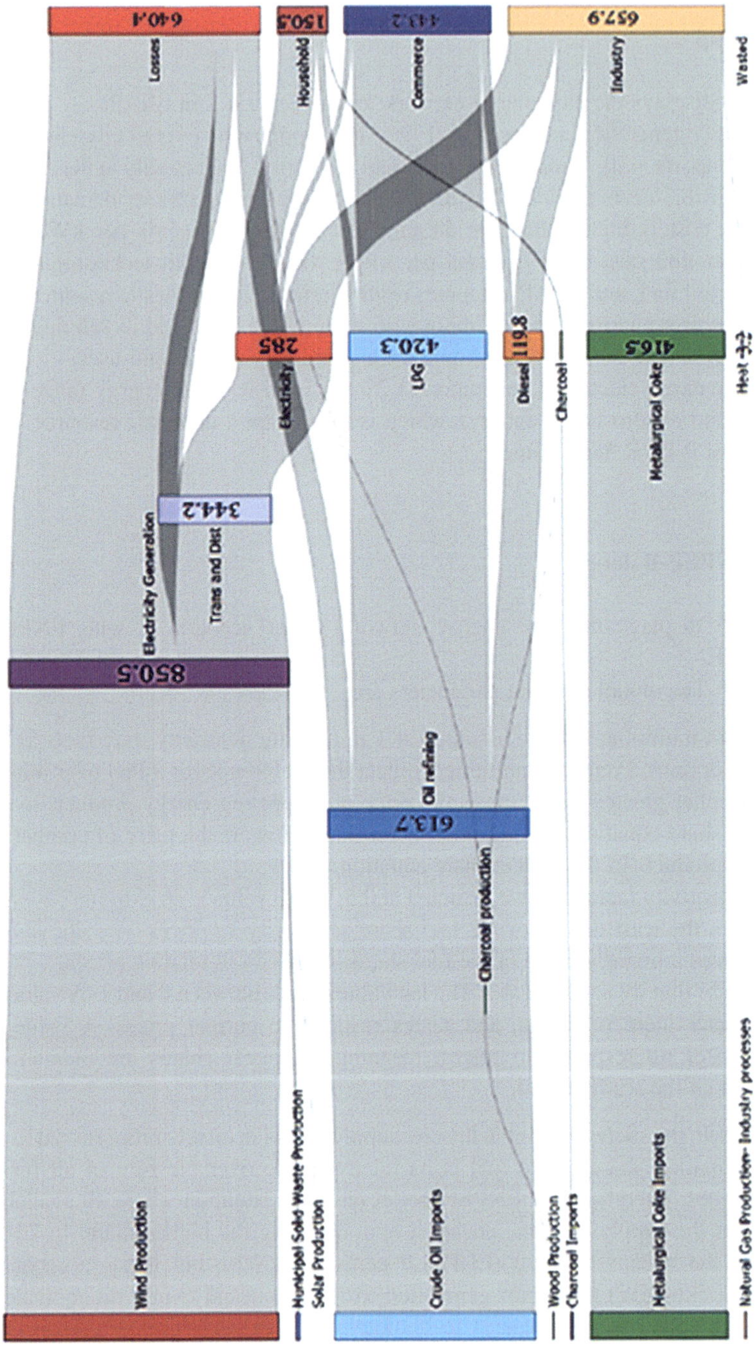

Fig. 9.4 LEAP-2017 City Energy System (in ktoe)

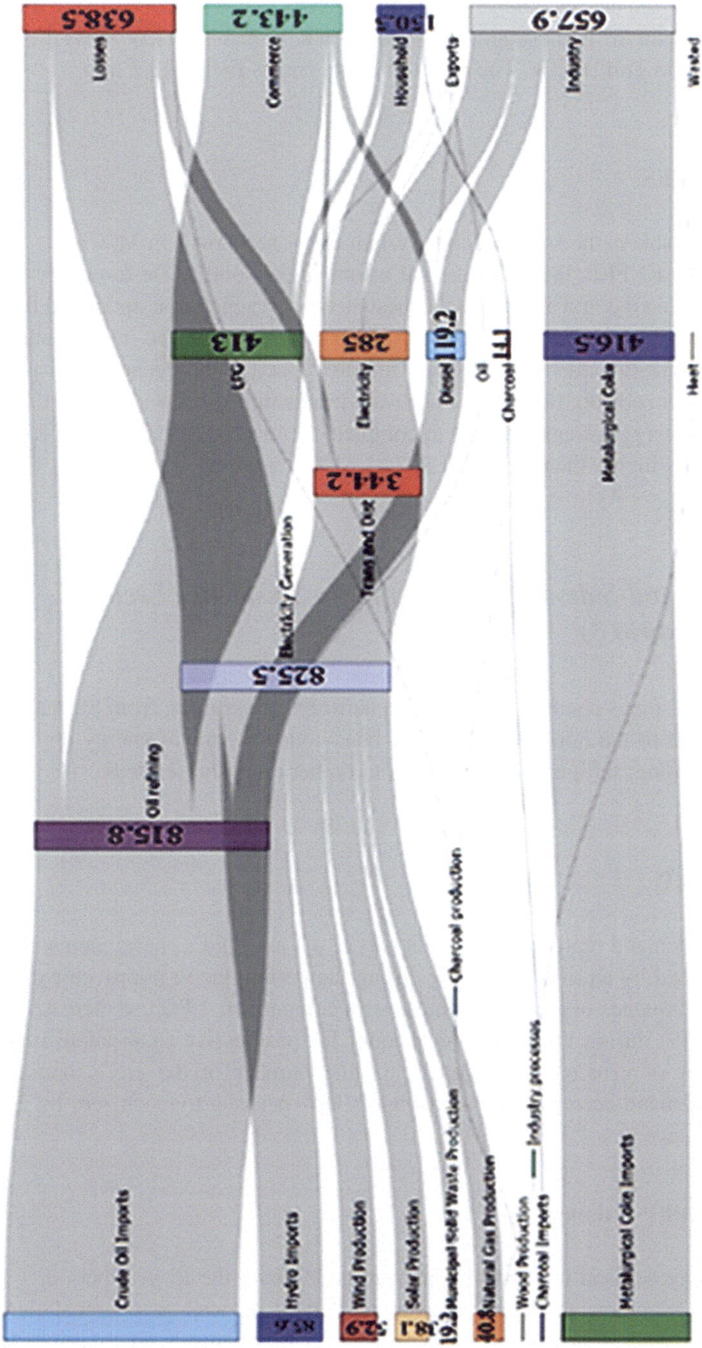

Fig. 9.5 ENPEP-2017 City Energy System (in ktoe)

with a unit generation cost of EUR 3.2 cents per kWh is the main supplier of the grid, followed by wind energy. Diesel generation decreases by 74% compared to the RES. Natural gas with the highest generation cost (EUR 17 cents per kWh) has the lowest contribution to grid supply. The overall system cost is EUR 943.2 mio.

4.2.3 MoCES

Figure 9.6 displays the future energy system under scenario 2 in MoCES.

Similar to ENPEP, the full potential of municipal solid waste-to-energy is integrated, as being the first and third-cheapest electricity generation options in the mix. It is followed by hydropower, diesel, and wind. Hydropower import enters the generation mix limited by its baseline import capacity, and because it is possible to limit power capacity (kW) in the software production window. About 11% of the city's wind energy potential is used to complete the mix. System cost is EUR 1088.8 mio, which is higher than the ENPEP-BALANCE resulting system cost.

4.3 Demand-Side-Management in the Residence Sector (Scenario 3)

Figure 9.7 displays result in LEAP of Scenario 3 that assumes, from Scenario 2, an improvement of the households' energy behaviour in terms of energy demand for cooling, lighting, refrigeration, cooking, and other domestic services.

4.3.1 LEAP

Electricity demand represents the main part of a household's energy demand, but it only increased by an average 1.52% during the period due to improvements in the energy performance of electric appliances. In comparison, LPG use increased by an average 5.5% during the period to account for progressive (interpolate function) replacement of charcoal and increase of the number of the city's households. Charcoal demand decreases by an average 19.8% per year to reach zero by 2030.

4.3.2 ENPEP-Balance

- Electricity demand increases by an average 2% over the 15-year period.
- LPG demand increases by an average 1.25%, as well as charcoal demand, because the presence in the energy mix of a fuel depends on its price, consequently charcoal quantities cannot reach zero as long as the fuel is priced. The

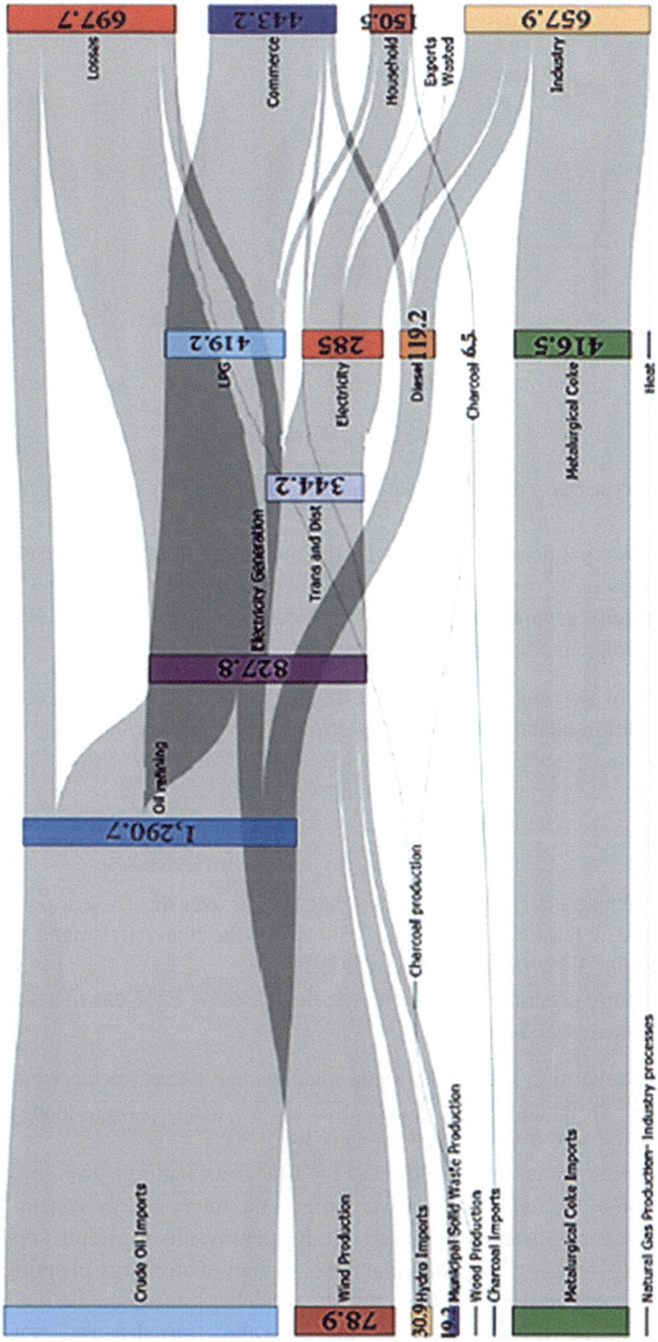

Fig. 9.6 MoCES-2017 City Energy System (in ktoe)

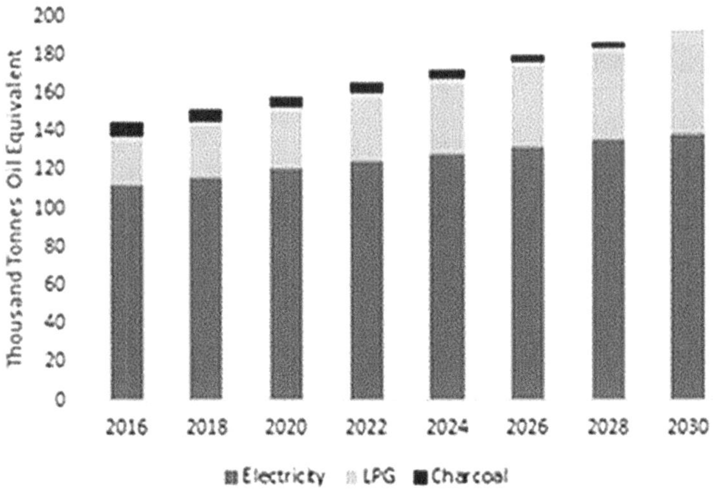

Fig. 9.7 Energy demand per fuel in the residence sector

growth of charcoal demand is driven by the increasing number of households, as for other fuels.

- Gross electricity generation, before losses, decreases by an average 1.4% due to decreasing demand compared to scenario 2.

As the model always runs to an equilibrium, all electricity generation technologies decrease generation quantities in the proportion of their market shares.

4.3.3 MoCES

- Electricity demand decreases by an average 0.17% over the 15-year period.
- LPG demand increases by an average 3.3%, while the charcoal demand decreases by an average 23.3% to reach zero by 2030.
- Gross electricity generation, before losses, decreases by 1.5% due to a decreasing demand compared to Scenario 2.

Compared to Scenario 2, the demand-side-management decreases the wind energy generation by 48% equivalent to a 572 MW wind power plant or the decommissioning of 310 MW diesel capacity by 2030.

In the following paragraphs, we discuss the four main findings from the models.

The selection of the modelling software affects the future energy system. From a single reference energy scenario, we derived different results regarding supply mix and system cost. Therefore, the claim that the outcomes of an energy planning model only depends on subjective considerations (the modeller preference) do not hold, as objective factors like the approach and the algorithm formulae also play out in the outcomes.

One hundred (100) per cent renewables supply mix such as in Scenario-2 LEAP does not guarantee the lowest system cost. Energy transition in the conditions of Scenario 2-LEAP is the most expensive for the city.

Dispatch of the city's generation technologies are considered to be the running cost rule, meaning technologies with the lowest generation cost (long-run generation cost) enter first the supply mix but, this does not guarantee the lowest electricity production cost. The ENPEP-BALANCE scenario-2 that integrates a share of all accessible technologies (with renewable and non-renewable energy resources) has the lowest cost compared to Scenario 2-LEAP (waste-to-energy and wind) and Scenario 2- MoCES (diesel, hydropower, waste-to-energy, and wind).

Accessible demand-side-management in the residence sector has a significant impact in the system's generation quantities. Measures such as retrofitting lamps, setting higher balance temperature point (BTP) in cooling can save up to the equivalent of 572 MW wind energy. The adoption of 100% LPG cooking saves 17 thousand tonnes oil equivalent of wood, which corresponds to a woodland area of 41 km2.

5 Conclusion

The study demonstrates that energy modelling software can integrate different dimensions of transition to energy sustainability at the city level, including the selection of electricity generation technologies, security of supply, and improved efficiency in energy use. Moreover, it particularly shows that different pathways exist to reach the same objective of transition to energy sustainability in the city network, and each produces different externalities. LEAP produces the most secure future energy scenario for the city by only using domestically available resources, but it is also the most expensive option. ENPEP displays the cheapest future energy scenario for the city, but it is also a less secure option as it continues to rely on imported sources that are cheaper than domestically available resources. MoCES displays results somewhere between security (more wind in supply) and affordability (presence of diesel).

Still, existent software also features limitations. These limitations include the abstraction of relevant energy parameters that affect the future system modelled by the software and the absence of flexibility in integrating different (agents) rationales on energy demand and supply options. Energy is a social good, as individual agents produce and consume its services; therefore, a relevant modelling approach should integrate the complexity of these agents' rationales to produce and/or to consume it. The cities' efforts to achieve energy sustainability (SDG-7) and urban liveability (SDG-11) by 2030 require relevant accounting methodologies and consistent model metrics (Grubler et al., 2012). MoCES, with an agent-based modelling approach, is a contribution to addressing both the relevance and consistency issues in energy planning models.

A. Annexes

Annex 1- Supplementary materials related to the SEA4cities survey on energy behavior of Dakar citizens.

The Table 9.6 summarizes the survey sample data.

Table 9.6 Survey strata

District	Residence buildings (a)	Sample size (b)	Number of residences surveyed (c)
Diamniadio	2165	327	368
Fann-point E Amitie	2128	326	360
Total	4293		728

[a]The number of residence buildings in a district is from ANSD (Diamniadio) and from the municipality of the district (Fann-Point E—Amitie)
[b]The sample size is computed with a confidence interval of 95% and $p = 0.5$
[c]The number of residence buildings that completed the survey questionnaire

Table 9.7 Average energy consumption

	Diamniadio	Fann—Point E—Amitie	City (Average)
Lighting (Wh) per day (a)	369,087.5	403,478.5	
Number of days (2016)	366	366	
Number of households	330	245	
Lighting (kWh per household)	409.4	602.7	506.1
Cooling (kWh per household) (b)	995.8	1311.2	1153.5
Refrigeration (kWh per household) (c)	1041.7	1553.3	1297.5
TV (kWh per household) (d)	35.8	110.7	73.2
Other appliances (kWh per household) (e)	29.7	42.4	36
Total (kWh per household)	2512.3	3620.3	3066.3
Total (toe per household)			0.26

[a]The row summarizes the total lighting energy of surveyed buildings per day. Lighting per household per day is computed with the following parameters from the questionnaire: type of lamps, lamp wattage, number of lamps in the building, and hours of operation per day
[b]The row is an average of energy consumed by space cooling appliances present in the buildings surveyed. The space cooling appliances from the completed questionnaires are air conditioners, celling and standing fans. Cooling energy per appliance per day is computed with the following parameters: appliance type, power, and average hours of operation per day in dry and rainy seasons
[c]The row is an average of energy consumed by fridges present in the buildings surveyed. Refrigeration energy is computed with following parameters: type of fridge, power, and hours of operation per day assumed to 24 hours for all fridges
[d]The row is an average of energy consumed by TVs present in the buildings surveyed. TV energy is computed with following parameters: type of TV, power, and hours of operation per day
Other appliances tracked in the survey are phones and laptops. The energy per appliance is computed with following parameters: number of appliances in the building, power and recharge hours per day (average value provided by the respondent)

Table 9.8 Grid Peak load (MW)

Month	Senegal [(a)]	Dakar [(b)]	in % of Dakar Peak
January	474	293.88	84.6%
February	476	295.12	85.0%
March	477	295.74	85.2%
April	489	303.18	87.3%
May	495	306.9	88.4%
June	528	327.36	94.3%
July	532	329.84	95.0%
August	544	337.28	97.1%
September	547	339.14	97.7%
October	560	347.2	100.0%
November	531	329.22	94.8%
December	512	317.44	91.4%

[a]Peak load data at the country level in 2016 is from the utility (SENELEC, 2017)

[b]Peak load data for the city assumes 62% of the country peak load. This figure was the share of electricity from the interconnected grid consumed in the city in 2016 (ANSD, 2018)

Dates of Survey

The survey in Diamniadio was completed between April and May 2019 (Dry season with average temperature of 30 degree Celsius).

The survey in Fann-Point E—Amitie was completed during November 2019 (End of the rainy season with average temperature of 25 degree Celsius). The Table 9.7 compiles results of the excel-based analysis of data collected during the survey.

Annex 2- Peak load data of the interconnected grid in 2016 (SENELEC, 2017). Table 9.8 provides peak load data in the interconnected grid for 2016.

References

Agence Nationale des Statistiques et de la Demographie. (2017). *Rapport global du Recensement général des Entreprises*. Agence Nationale des Statistiques et de la Demographie.

ANSD. (2018). *Dakar: situation economique et sociale regionale 2016*. Agence Nationale des Statistiques et de la Demographie. Consulté le November 23 2019, sur http://www.ansd.sn/ressources/ses/SES-Dakar-2016.pdf

ArgonneNational Laboratory. (2019). *ENPEP-BALANCE software*. Lemont. Retrieved October 23, 2019, from ww.anl.gov

CRSE. (2017). *Bilan des activites de SENELEC 2014–2016*. Commission de Regulation du Secteur de l'Energie. Consulté le June 10, 2019, sur https://www.crse.sn/sites/default/files/2017-04/RevisionTarifaire-2016b_0.pdf

Fall, A. (2020). *Techno-economic assessment of renewable's energy potential of the city:Examples of solar photovoltaic, waste-to-enery and wind energy*. Unpublished.

Fall, A., Ndiaye, C. A., & Mbodji, C. A. (2020). *Democratization of energy planning: On a new planning tool tailored to the needs of developing countries*. Unpublished.

Grubler, A., Bai, X., Buettner, T., Dhakal, S., Fisk, D. J., Ichinose, T., ... Weisz, H. (2012). Chapter 18–urban energy systems. In A. Grubler (Ed.), *Global energy assessment–toward a sustainable future* (pp. 1307–1400). Cambridge University Press. Retrieved from http://pure.iiasa.ac.at/id/eprint/10067/1/GEA%20Chapter%2018%20Urban%20Energy%20Systems.pdf
Ministere en charge des Energies. (2018). *Rapport SIE Senegal 2016*. Ministere en charge Energies.
Ringkjøb, H.-K., Haugan, M., & P., & Solbrekke, I. M. (2018). A review of modelling tools for energy and electricity systems with large shares of variable renewables. *Renewable and Sustainable Energy Reviews*, 440–459. https://doi.org/10.1016/j.rser.2018.08.002
SENELEC. (2017). *Rapport annuel*. SENELEC.
Statista. (2020). *Monthly exchange rate of the U.S. dollar to euro from January 2014 to October 2020*. Hamburg. Retrieved January 7, 2020, from https://www.statista.com/statistics/247846/exchange-rate-of-the-euro-against-the-us-dollar-monthly-development/
StockholmEnvironment Institute. (2020). LEAP. Stockholm, Sweden. Consulté le January 2020, Sur LEAP.: https://www.energycommunity.org
World Bank Group. (2019). *Urban Population Growth (annual %)*. Retrieved January 11, 2020, from The World Bank Data: https://data.worldbank.org/

Chapter 10
Management of Intermittent Solar and Wind Energy Resources: Storage and Grid Stabilization

W. M. Nkounga, M. F. Ndiaye, and M. L. Ndiaye

Abstract The chapter documents options for management of the intermittency of solar and wind energy resources, with the aim of supporting transition to energy sustainability with these resources. It explores different techniques for creating storage in high power and high energy systems. We review indicators to support the decision on the selection of these storage options combined or not to grid management strategies. Our results show that flywheel is more appropriate in short-term high power storage given its low investment cost and its power density per cubic metre. For long-term energy storage, still considering the investment cost and power density per cubic metre, hydrogen, and hydraulic pumping are the best options. The smart management of storage options can significantly reduce the impact of solar and wind resources intermittency on the stability of the grid.

Keywords Energy · Renewable energy · Intermittency · Storage · Network management

1 Introduction

Renewable energy accounts for 26% of the world's electricity production (IEA, 2020). Renewables in electricity generation provide environmental and economic benefits, which include the reduction of greenhouse gas emissions, mitigation of climate variability, and energy independence.

W. M. Nkounga (✉)
École Supérieure Polytechnique de Dakar, Dakar, Sénégal

Polytech Annecy-Chambery, Annecy, France
e-mail: wellymagloire.nkounga@ucad.edu.sn; willy-magloire.nkounga@univ-smb.fr

M. F. Ndiaye · M. L. Ndiaye
École Supérieure Polytechnique de Dakar, Dakar, Sénégal
e-mail: mouhamadouf.ndiaye@ucad.edu.sn; mamadoulamine.ndiaye@ucad.edu.sn

A. Fall, R. Haas (eds.), *Sustainable Energy Access for Communities*,
https://doi.org/10.1007/978-3-030-68410-5_10

However, the intermittency of some renewable energy resources, such as solar and wind energy, is a major concern when the generation systems are connected to the grid. Therefore, several techniques are proposed in the scientific literature to address the issue of managing intermittent solar and wind energy resources: short, medium, and long-term forecasting of resource availability (Nkounga et al., 2018; Javed et al., 2020), geographical dispersion (spatial diversification) of production units (Liu et al., 2020), storage (Chatzivasileiadi et al., 2013), grid interconnection, and hybrid systems. In this chapter, we focus on storage and network interconnection techniques.

Energy storage options are numerous and include hydraulic pumping, fuel cell, flywheel, and the combinations battery/hydraulic pump, battery/supercapacitor, battery/fuel cell, battery/flywheel, and battery/flywheel/supercapacitor (Javed et al., 2020). However, a number of problems are prevalent during the operation of these storage systems: low predictability over time, lack of suitable location, high investment costs, low level of autonomy, short discharge time. These problems constitute additional obstacles to the integration of wind and solar energy systems into electricity networks beyond investment in power capacities.

In terms of capacities for electricity generation, solar photovoltaic and wind energy are among the most advanced renewable energy technologies that have been integrated into the main electricity grid in several regions of the world (Al-Shetwi et al., 2020). Connecting these systems to the centralized grid raises issues that include voltage fluctuation, reactive power, grid overload and harmonics distortions (Benzohra et al., 2020). These issues pose significant challenges in terms of power factor, storage management, energy forecasting and planning (Shafiullaha et al., 2018). These issues also raise the following question: How could solar and wind energy systems be successfully integrated into power grids over the long term and at low cost, while optimizing grid stability?

The objective of this study is to propose effective measures for managing the intermittency of solar and wind energy resources, the implementation of which would constitute a contribution to achieving the objectives of Sustainable Development Goal number 7 (SDG-7). The chapter includes five sections: Following the introduction, Sect. 2 describes the study's methodological approach. Section 3 presents the tools for the management of storage techniques, which include their configurations and the conditions of implementation for improved efficiency. Section 4 compiles discussions and recommendations. Finally, the conclusion and perspectives are presented in Sect. 5.

2 Methodological Approach

The study methodology is a systematic literature review followed by the definition of a protocol to address issues associated with intermittency of solar photovoltaic and wind energy in the literature. The protocol includes the following steps:

- Classification and comparative analysis of power storage techniques.
- Classification and comparative analysis of energy storage techniques.
- Classification of storage techniques according to the planning horizon considered.
- Identification of storage techniques considering investment costs, charging time, discharge time, required surface area, density per cubic metre related to each technique of storage for both power and energy storage.
- Analysis of grid connection as backup option based on stochastic, deterministic, and hybrid management strategies (simultaneous use of stochastic and deterministic management techniques) suitable for solar photovoltaic and wind energy.

3 Results

Figure 10.1 displays a comparison of investment costs for different techniques of power storage. The blue and red bars represent the minimum and average investment costs for each type of storage, respectively. For power storage, hydraulic pumping, compressed air, hydrogen, and batteries have a relatively high investment cost per kilowatt compared to other techniques. Flywheel, magnetic conductivity and supercapacitor storage techniques have a lower investment cost per kilowatt; their minimum investment costs are between EUR 100 and 400 per kW.

Flywheel and magnetic conductivity storage systems have similar investment costs that are relatively low, which explains why they are the techniques most widely used for power storage. Conventional batteries, in particular lead, nickel, lithium, and zinc-air batteries, are also technologies frequently used in storing power, despite

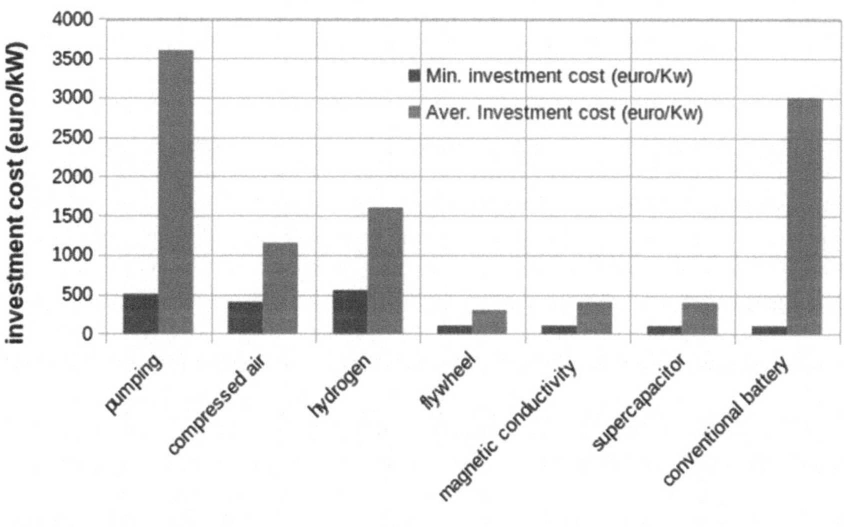

Fig. 10.1 Investment costs of power storage systems

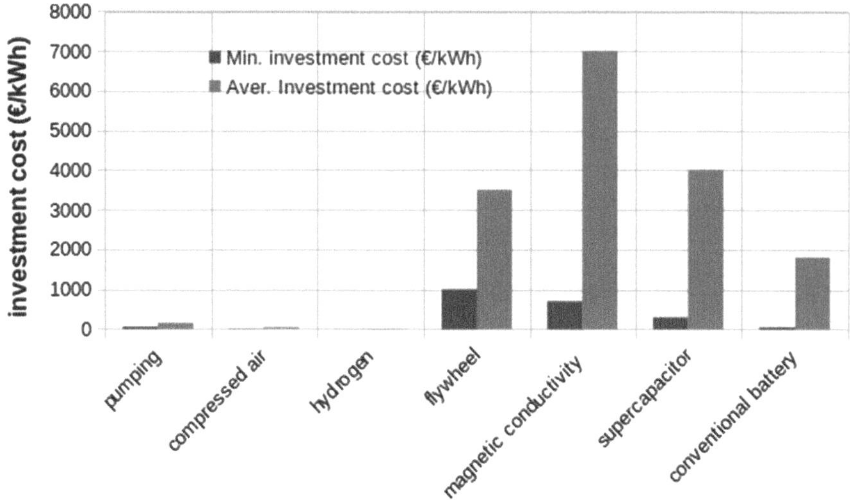

Fig. 10.2 investment costs of energy storage systems

their prohibitive costs. The cost of storage batteries varies according to the technology. The average cost of EUR 3000 per kW corresponds to lithium-ion technology and the lowest cost to sodium-nickel (NaNiCl) technology.

Figure 10.2 shows a comparison of investment costs for energy storage. The minimum investment cost is shown in blue, and the average investment cost is in red.

For energy storage, hydraulic pumping, compressed air, and hydrogen feature the lowest investment costs for long-term energy storage. The flywheel, magnetic conductivity and supercapacitor have relatively high investment costs. Pumping requires an investment of EUR 60–150 per kWh, compressed air requires an investment of EUR 10–40 per kWh, and hydrogen requires an investment of EUR 1–15 per kWh. Conventional battery, here again, stands out with a cost that is relatively high and utilization that is frequent in energy systems. The battery investment cost depends on the technology; the minimum cost of EUR 50 per kWh corresponds to lead-acid batteries, and the average cost of EUR 1800 per kWh corresponds to lithium batteries. The latter feature the advantages of having much higher efficiency and autonomy period. Table 10.1 provides a classification of storage techniques according to discharge time, lifetime, self-discharge time and recharge time.

The discharge time of storage options varies from 1 second to several days. The technologies featuring the lowest discharge times are supercapacitor, magnetic conductivity, and flywheel, with a limited duration of a few milliseconds for magnetic conductivity and supercapacitor, and a maximum duration of around 15 minutes for the flywheel.

Considering the service life of the storage techniques, hydraulic pumping can operate for one hundred years (one century). Conversely, hydrogen and battery

Table 10.1 Durability indicators of storage techniques (Chatzivasileiadi, Ampatzi, & IanKnight, 2013)

Storage	Discharge time	Service life (years)	Discharge time (% day)	Recharge time
Pumping	h-jr	50–100	0	Min-h
Compressed air	h-jr	25–40	0	Min-h
Hydrogen	s-jr	5–15	0.5–2	Instantaneous
Flywheel	15 s-15 min	Sup à 20	20–100	14 min
Magnetic conductivity	Ms-1 h	Sup à 20	10–15	Min
Supercapacitor	Ms-1 h	Sup à 20	2–40	s-min
Conventional battery	s-h	3–30	0–20	Min-16 h

Table 10.2 Indicators of energy storage (Chatzivasileiadi, Ampatzi, & IanKnight, 2013)

Storage	Energy (kWh)	Specific energy (Wh/kg)	Density (kWh/m3)	Required space (Wh/m2)
Pumping	200–5000	0.5–1.5	0.2–2	20
Compressed air	200–1000	30–60	12	100–28
Hydrogen	10,000	33.33	600	50–60
Flywheel	10–5000	5–130	20–80	280–610
Magnetic conductivity	1–100	0.5–5	6	930–26,000
Supercapacitor	0.001–10	0.1–15	10–20	430
Conventional battery	0.01–100,000	30–400	20–800	20–60

storage have lifetimes comprised between 5 to 15 years and 3 to 30 years, respectively.

The recharge time in the fifth column of Table 10.1 shows instantaneous recharging for hydrogen storage. This makes it the fastest option. Conversely, conventional batteries can take up to 16 hours for a full recharge.

Table 10.2 provides a classification of storage options, considering production, density, and space requirements of the facilities.

Among the storage options in Table 10.2, magnetic conductivity requires the widest space. A production of 1–100 kWh requires 930 to 26,000 Wh/m^2, compared to 20 Wh/m^2 for 200 to 5000 kWh of a hydraulic pumping system. After hydraulic pumping, battery, and hydrogen are the least space-consuming storage systems. This characteristic provides them with value addition in long-term energy storage.

Table 10.3 provides a classification of storage systems, considering the power and density of the systems.

Magnetic conductivity and supercapacitor have a relatively high density of 4000 to 120,000 kW/m^3 for 0.01 to 1 megawatt power storage. Compressed air and hydraulic pumping systems have a relatively low-density compared to other

Table 10.3 Indicators of power storage (Mahmoud et al., 2020)

Storage	Power (MW)	Density (kW/m3)
Pumping	100–5000	0.1–0.2
Compressed air	100–300	0.2–0.6
Hydrogen	49	0.2–20
Flywheel	0.01–1	2600
Magnetic conductivity	0.01–1	40,000-120,000
Supercapacitor	0.01–1	40,000-120,000
Conventional battery	0.5–69	0.5–10,000

Table 10.4 Energy management techniques (Bukar & Tan, 2019)

Objective functions	Methods	Constraints
Minimizing energy cost	Deterministic	Battery, grid capacity, reactive power
Minimizing costs	Deterministic	Solar and wind power, state of charge (SOC) of storage units
Minimize network losses and costs	Stochastic	Battery characteristics, solar, and wind energy resources
Minimize annual costs, and energy losses	Stochastic	State of charge (SOC), photovoltaic power, grid, power factor
Minimize maintenance cost and power supply losses	Stochastic	SOC, energy efficiency
Minimize system cost and energy surplus	Stochastic	Tilt, number of solar panels or wind turbines, storage capacity

techniques of storage. For a system of 100 to 5000 MW, a density of 0.1 to 0.2 kW/m^3 is required for hydraulic pumping and for a system of 100 to 300 MW, a density of 0.2 to 0.6 kW/m^3 is required for compressed air.

Table 10.4 presents techniques often used in managing the integration of solar and wind energy systems connected into the grid with storage. The management strategies are based on smart monitoring and control protocols, which are associated with constraints and objective functions.

Deterministic methods are management techniques based on linear programming, with numerical, analytical, iterative methods, and probabilistic computations, and graphic construction. Deterministic methods relate to genetic algorithms, particle swarm optimization, bee colonies, simulated annealing, biogeographic optimization, and imperialist competitive algorithms (Bukar & Tan, 2019). These techniques are used in the management or energy systems with storage and with or without grid connection. The appropriateness of these management techniques varies depending on the objective constraints and functions. In the effective management of intermittent solar photovoltaic and wind energy resources, each of these methods can apply objectively, depending on the objective functions. Table 10.4 presents some of these techniques considering the objective function and some constraints.

The hybrid (stochastic-deterministic) approach can also apply on some occasions. The elements presented in Table 10.4 for the management of the intermittency of solar and wind energy resources are structured based on established algorithmic protocols. The constraints represented are necessary to satisfy throughout the control process. The battery state of charge (SOC), energy demand, and the availability of solar and wind resources are basic indicators, which condition intervention in the network. Minimisation of costs, reactive power, and power factor are the constraints associated with the grid; the control strategy is successful when the grid constraints, resource constraints, and demand constraints are satisfied simultaneously.

4 Discussion of Results

The results observed in the previous section demonstrate that fluctuations in energy networks can be controlled despite the intermittency of solar and wind energy resources in the network. In the short term, high power systems can be associated with flywheel, magnetic conductivity and supercapacitors storage techniques. This observation complies with the recommendations of Chatzivasileiadi et al. (2013), which propose the use of flywheel (rotational energy), super magnetic conductivity and supercapacitors (in electrostatic form) storage technologies in the short term. These options should be prioritized in planning additional renewable energy capacities as a contribution to Sustainable Development Goal 7 (SDG-7). According to the European Patent Office quoting the International Energy Agency, between 189 and 305 GW of energy storage capacity will be needed by 2050 to mitigate the impact of connecting intermittent renewable energy power systems in energy networks (European Patent Office, n.d.).

With regard to long-term energy storage, hydraulic pumping offers interesting characteristics, especially for countries in the sub-Saharan Africa region, due to its relatively low investment cost and availability of the resource. This observation complies with the recommendations of Javed et al. (2020) and those of Mahmoud et al. (2020) for using this technology. This observation also explains the increasing popularity of hydraulic pumping in recent renewable energy generation systems (see Fig. 10.3). According to the International Hydropower Agency, 3.2 GW of hydraulic pumping storage was added between 2017 and 2018 (International Hydropower Association, 2018). According to Mahmoud et al. (2020), mechanical storage has an advantage in environmental impact, cost, and sustainability. Thus, compressed air and hydraulic pumping are relevant storage options to address the concerns that raise electricity generation with intermittent solar and wind energy resources in the region. Currently, only two power plants with compressed air storage are operational worldwide (110 MW in the USA and 290 MW in Germany), compared with about a hundred power plants associated with hydraulic pumping storage.

On connection to the grid as backup, the appropriate management strategy should consider the objective functions and the corresponding constraints, as proposed in Table 10.4. Control and command strategies defined considering this approach make it possible to anticipate potential energy deficit or surplus while addressing issues related to intermittencies such as load balancing, peak demand alleviation, cost minimisation, and energy efficiency (Mahmoud et al., 2020).

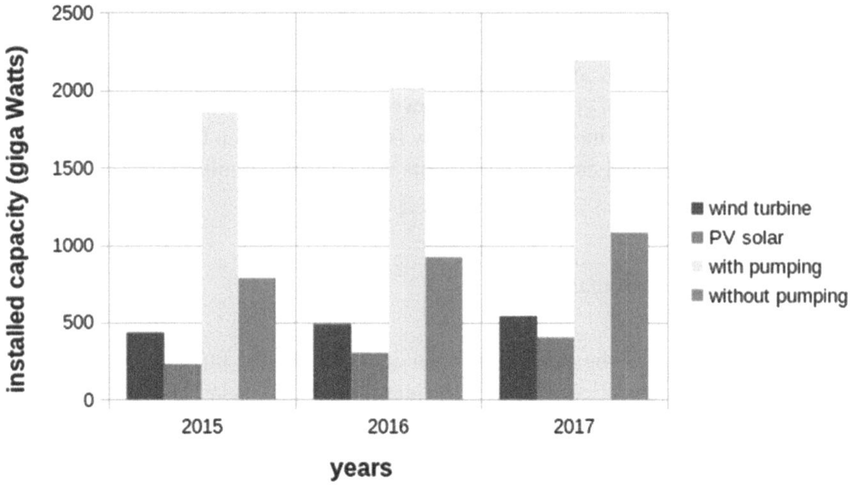

Fig. 10.3 Capacity of storage systems installed worldwide (Javed et al., 2020)

5 Conclusion

In this chapter, we explore different storage systems that could contribute to addressing the issues associated with the intermittency of solar photovoltaic and wind energy resources connected to the grid. The analysis of storage techniques considers, among other parameters, their investment costs, their durability, density, and space required.

The study shows that parameters such as power and energy density, available space, service life, charge, and discharge duration are key factors in the selection of the appropriate storage technology. In the short term, taking into account investment costs and power density per cubic meter, flywheel is the best option for power storage. For long-term energy storage, when only considering the investment cost, hydrogen appears as a good option. However, when the required surface area and power density per cubic metre are taken into account, hydraulic pumping is a better option.

The inclusion of these indicators in the selection of a storage system contributes to improving the efficiency and attractiveness of renewable energy systems, because it reduces the investment risks and uncertainties associated with grid stability due to the intermittency of these resources. However, the value of relevant indicators is context-specific. Therefore, the selection of renewable energy resources to feed into the grid and storage options should consider strategic management techniques and context-specific parameters such as available surface areas and desired scale of production.

This study is limited to the exploration of storage techniques in the management of intermittent solar and wind energy resources connected to the electricity grid. The extension of its results considering the integration of environmental and social factors such as the forecast of solar and wind energy potential and energy demands in a specific context could be envisioned after an adaptation of the protocol presented here. Indeed, the integration in the analysis of energy resources forecast methodologies can support selections of the appropriate storage techniques and grid management strategies for different communities.

References

Al-Shetwi, Q., Hannan, M. A., Jern, K. P., Mansur, M., & Mahlia, T. M. (2020). Grid-connected renewable energy sources: Review of the recent integration requirements and control methods. *Journal of Cleaner Production*. https://doi.org/10.1016/j.jclepro.2019.119831

Benzohra, O., Echcharqaouy, S. S., Fraija, F., & Saifaoui, D. (2020). Integrating wind energy into the power grid: Impact and solutions. *Materials Today Proceedings, 30*, 987–992. https://doi.org/10.1016/j.matpr.2020.04.363

Bukar, A. L., & Tan, C. W. (2019). A review on stand-alone photovoltaic-wind energy system with fuel cell: System optimization and energy management strategy. *Journal of Cleaner Production*, 73–88. https://doi.org/10.1016/j.jclepro.2019.02.228

Chatzivasileiadi, A., Ampatzi, E., & IanKnight. (2013). Characteristics of electrical energy storage technologies and their applications in buildings. *Renewable and Sustainable Energy Reviews, 25*, 814–830. https://doi.org/10.1016/j.rser.2013.05.023

European Patent Office. (n.d.). *Energy storage: Added stability for electrical grids*. Retrieved November 2020, from https://www.epo.org/news-events/in-focus/sustainable-technologies/storage.html

IEA. (2020). *World energy outlook 2020*. International Energy Agency.

International Hydropower Association. (2018). *2018 Hydropower status report shows record rise in clean electricity*. International Hydropower Association. Retrieved November 23, 2020, from https://www.hydropower.org/news/2018-hydropower-status-report-shows-record-rise-in-clean-electricity

Javed, M. S., Ma, T., Jurasz, J., & Amin, M. Y. (2020). Solar and wind power generation systems with pumped hydro storage: Review and future perspectives. *Renewable Energy, 148*, 176–192. https://doi.org/10.1016/j.renene.2019.11.157

Liu, B., Lund, J. R., Liao, S., Jin, X., Liu, L., & Cheng, C. (2020). Optimal power peak shaving using hydropower to complement wind and solar power uncertainty. *Energy Conversion and Management, 209*. https://doi.org/10.1016/j.enconman.2020.112628

Mahmoud, M., Ramadan, M., Olabi, A.-G., Pullen, K., & Nahera, S. (2020). A review of mechanical energy storage systems combined with wind and solar applications. *Energy Conversion and Management*, 210. https://doi.org/10.1016/j.enconman.2020.112670

Nkounga, W. M., N'diaye, M. F., N'diaye, M. L., Cisse, O., Bop, M., & Sioutas, A. (2018). Short-term forecasting for solar irradiation based on the multi-layer neural network with the Levenberg-Marquardt algorithm and meteorological data: Application to the Gandon site in Senegal. In *7th International Conference on Renewable Energy Research and Applications (ICRERA)* (pp. 869–874). Paris. https://doi.org/10.1109/ICRERA.2018.8566850

Shafiullaha, G. M., Arif, M. T., & Oob, A. M. (2018). Mitigation strategies to minimize potential technical challenges of renewable energy integration. *Sustainable Energy Technologies and Assessments*, 24–42. https://doi.org/10.1016/j.seta.2017.10.008

Chapter 11
Innovation for the Better: How Renewable Energy Technologies Improve Living Standards

Abdoulaye Thiaw and S. Toure

Abstract Energy has always been of paramount importance for human societies. It played an instrumental role in the progress of nations. Throughout history, communities have sought to control and manage energy resources through technology. This chapter presents examples of innovation in energy technologies and services that aimed to improve living standards in local communities. The study exemplifies communities in the districts of Pikine and Guédiawaye, with solar photovoltaic systems to improve access in high schools. Our ambition is to learn from motivations, experience and people's testimonials, the determinants of a successful energy transition in local communities.

Keyword Solar photovoltaic · Sizing · Energy Cost · Education

1 Introduction

Energy and human progress feature both shades of a metaphysical and physical question. Human behaviour is affected by the characteristics of accessible energy sources. These behaviours range from the exploitation of fossil fuels that helped nations prosper and cause global changes to the use of locally available renewable energy resources that open a new era of technological innovation. In between lies the conflicting forces of routine with supply from conventional energy resources that the recent COVID-19 pandemic has made exceptionally cheap and the aspirations of citizens for new standards of living that preserve the common goods such as climate and biodiversity.

A. Thiaw (✉)
Universite Cheikh Anta Diop Dakar, Dakar, Senegal

S. Toure
Istanbul, Turkey
e-mail: Saliou.toure@undp.org

A. Fall, R. Haas (eds.), *Sustainable Energy Access for Communities*,
https://doi.org/10.1007/978-3-030-68410-5_11

Energy sustainability is at the confluence of environment friendliness, economic accessibility and social acceptance. For a long time, the analysis of energy-related issues focused on one or the other of these dimensions without exploring their nexus regions. An example is an approach in modelling energy systems, whether computational equilibrium or scenario simulation, which determines the cost of the future sustainable energy system but overlooks its environment and social impact.

In this chapter, we document the case of the energy-education nexus and learn from examples the drivers of success in innovative energy solutions to improve the living standards of communities in sub-Saharan Africa.

2 Methodological Approach

The study methodology consists in case studies, along with data collection and analysis. Data of the case studies exemplified are from the inter-municipal cooperation project jointly implemented by the municipalities of Pikine and Guediawaye in Senegal. The project targets high schools with solar power systems to improve education facilities.

Within the project implementation, two pilot systems have been installed in Limamoulaye (45 kW) and Thiaroye (30 kW) high schools to connect new educational materials such as desktops while reducing the electricity bill of the municipalities. A photovoltaic monitoring system via inverters was also installed in order to centralize the monitoring data and possibly access it via software. The monitoring system collects information on the following indicators:

- Electricity production per period
- Electricity consumption per period
- Indicators of operations in the remote maintenance of the systems
- Detection of deficiencies in the solar photovoltaic systems

The data collection covered a period of 12 months from January to December 2019 and was carried out by retrieving detailed inverter parameters, alarm production data and historical data messages recorded in real time. The data can be viewed remotely since the recorders have an integrated server that enables monitoring of the system, which contributes to limit production losses and improve the performance of solar systems (Bressan, 2014).

3 Results of the Analysis

3.1 Electricity Supply from the Solar Photovoltaic Systems

Figures 11.1 and 11.2 display the production curve of the solar systems and the energy consumption curve of each of the two high schools.

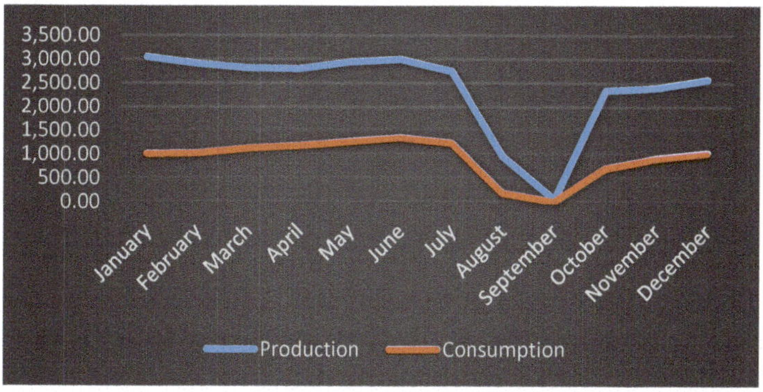

Fig. 11.1 Solar production versus Energy consumption for Limamoulaye High School

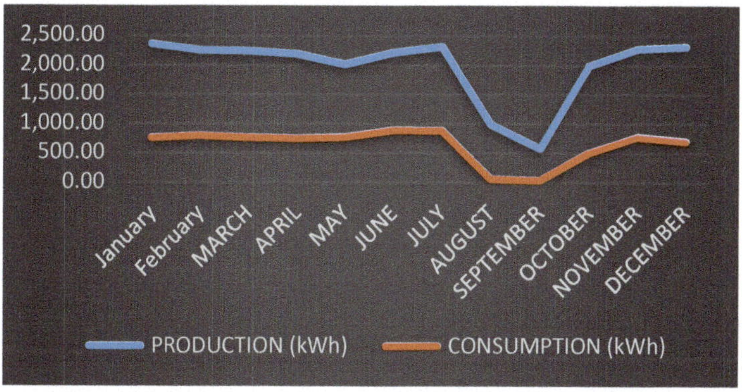

Fig. 11.2 Solar production versus Energy consumption for Thiaroye High School

The analysis of the Figure shows a regular energy production by solar systems between January and July, with an average of 2386 kWh per month. We can see a decline of both curves between the end of July and September due to the reduction of school activities; the academic year finishing at the end of July. This period also coincides with the rainy season, when clouds affect solar irradiation. This demonstrates the adequacy of solar photovoltaic systems with schools' activities in Senegal. From October, the solar system progressively increases production due to an increase in solar irradiation. October also coincides with the start of a new academic year.

Consumption over the period followed the same path, with demand being constantly below the production capacity of the system at an average of 919 kWh per month. This demonstrates an oversizing of the solar system.

The analysis of Fig. 11.2 shows patterns similar to Fig. 11.1. The average electricity production of the solar system during the period January-July is 1965 kWh, and the average consumption of school activities during the same period is 796 kWh.

3.2 Electricity Supply from the Interconnected Grid

Table 11.1 displays the quantities and costs of electricity supplied by the interconnected grid to Limamoulaye High School during the period from January to December 2018, 1 year before the installation of the solar photovoltaic system.

Table 11.2 displays the quantities and costs of electricity supplied by the interconnected grid to Thiaroye High School during the period from January to December 2018, 1 year before the installation of the solar photovoltaic system.

The economic value of the solar photovoltaic system is measured by comparing the investment cost, and savings on periodic energy bills received the year before the system installation from the national utility, which is also the manager of the grid.

Table 11.1 Electricity supply from the grid (Limamoulaye High School)

Month	Cons.	Cost (FCFA 110 per kWh)	Cost (EUR 0.17 per kWh)
January	1769	194,590	296.6
February	1995	219,450	334.5
March	1772	194,920	297.1
April	1829	201,190	306.7
May	1790	196,900	300.2
June	1772	194,920	297.1
July	1521	167,310	255.1
August	575	63,250	96.4
September	750	82,500	125.8
October	1189	130,790	199.4
November	911	100,210	152.8
December	1482	163,020	248.5
Total		1,909,050	2910

Table 11.2 Electricity supply from the grid (Thiaroye High School)

Month	Cons.	Cost (FCFA 110 per kWh)	Cost (EUR 0.17 per kWh)
January	1269	164,970	251.5
February	1195	155,350	236.8
March	1372	178,360	271.9
April	1229	159,770	243.6
May	1190	154,700	235.8
June	1372	178,360	271.9
July	1421	184,730	281.6
August	375	48,750	74.3
September	450	58,500	89.2
October	1100	143,000	218
November	850	110,500	168.4
December	1350	175,500	267.5
Total		1,712,490	2610.7

With an average cost of EUR 3 per Watt installed at Limamoulaye high school, the system returns a cost-saving of EUR 2910 per annum. This corresponds to a payback period of forty-six (46) years and a net present value of EUR -111,674.35. With an average cost of EUR 3 per Watt installed in Thiaroye high school, the system returns a cost-saving of EUR 2610.7 per annum. This corresponds to a payback period of thirty-four (34) years and a net present value of EUR – 69,073.4.

4 Discussion of Results

Findings in the previous section actually reveal a number of issues related to the demonstration of renewable energy technologies in the local communities of developing countries, besides their impact on living standards.

The problem relates to the solar photovoltaic system sizing. In both schools, it appears that the capacity of production installed is higher than the demand. The problem derives from a combination of three factors: (1) models of acquisition of the renewable energy technology, (2) focus on the restrictive dimensions of energy sustainability, and (3) lack of capacity of local installers.

The first dimension of the problem relates to the donor-led initiatives structure. Usually, pilot projects to estimate the relevance of renewable energy technologies in local communities of sub-Saharan Africa are financed with donations from non-governmental organizations or multilateral international cooperation agencies. The absence of profitability criteria in the transaction creates a perception of gratuity, which overlooks the economic dimension of the sustainability agenda and compromises replicability. The municipality accounts for savings from bills without realizing that the amounts saved are at an initial cost that is much higher. It clearly appears that financial figures computed for both installations, meaning payback period and net present value, are unfavourable, and the model could not be replicated with technologies that are acquired from private suppliers.

In addition to energy bills, the system definitely reduces CO2 emissions. Electricity generation from the Senegal utility emits an average 0.56 tonnes CO2-eq per MWh (UNFCCC, 2017). Therefore, a replacement by solar photovoltaic generation avoids an equivalent amount of emissions. The solar photovoltaic systems also feature a positive social impact; the interconnected grid supply is erratic, and power cuts are frequent. Therefore, self-production with rooftop solar photovoltaic systems ensures control over the electricity supply of the school and consequently powers some academic activities such as learning with computers.

The lack of qualified technical resources in local communities, especially in remote rural areas of sub-Saharan Africa, is another drawback in the energy transition agenda. Sizing is a necessary step in the installation of decentralized energy technologies. However, when capacities are not available, the installation of renewable energy technologies only consists of connecting wires and devices. Figures 11.1 and 11.2 with production versus consumption curves show that the solar photovoltaic systems available in these municipalities could power, at least, two additional buildings with similar consumption patterns. This would correspond to division by a factor of 2 of the payback period and a multiplication of the net present value by a factor of 1.4.

5 Conclusion

It is indisputable that a transition towards the sustainability of energy systems features positive values in sub-Saharan Africa. Firstly, the renewable energy resources, including solar irradiation and insulation hours, is readily available. Secondly, the business-as-usual energy production processes use fossil fuels that emit greenhouse gases such as carbon dioxide (CO2), methane (CH4), and nitrous oxides (NOx). Therefore, a replacement with renewable energy technologies such as solar photovoltaic systems avoids these emissions. Thirdly, the obsolescence of the grid infrastructure in many sub-Saharan Africa countries or sometimes the prohibitive costs of grid extension to rural communities make supply with decentralized systems more secure. However, the business model of acquisition, planning and installation makes renewable energy technologies unsustainable. This contributes to the still common perception of the energy transition agenda as expensive compared to the routine of supply with the interconnected grid.

From the two examples presented in this chapter, it appears that some factors should be investigated in priority, which includes the technology acquisition model (sale versus donation) and the technical capacity of human resources available in local communities for the installation of technologies. Both require prior attention before the definition of the sustainable energy agenda. Otherwise, the process of transition to energy sustainability will continue to stagnate from pilot installations of different renewable energy technologies without viable replication models.

References

Bressan, M. (2014). *éveloppement d'un outil de supervision et de contrôle pour une installation solairephotovoltaïque*. Universite de Perpignan. Consulté le June 2020, sur https://tel.archives-ouvertes.fr/tel-01068025/document

UNFCCC. (2017). *Standardized baseline: Grid emission factor for west African power Pool*. United Nations Framework Convention on Climate Change.

Chapter 12
Electricity Consumption in Working-Class Districts: Case Studies of Grand-Yoff and Grand-Dakar

S. Sow, C. M. Kebe, and A. Ndiaye

Abstract The chapter aims to explore the determinants of electricity consumption in working-class districts of Senegal cities. Grand-Yoff and Grand-Dakar are among the 19 district municipalities of the city of Dakar. The study collected data on different parameters related to electricity consumption that includes the number and type of plugging appliances. The methodology consists of a comparison between real electricity consumption of residential buildings and estimations based on models from the literature, which is followed by an analysis of the impact that a number of socio-economic parameters can have on the annual electricity consumption. Our results show that estimations can significantly differ from data collected in the field. They also show that context-specific, social and economic parameters affect electricity consumption in various magnitudes. Therefore, the planning of electricity production systems such as community grids or other decentralized systems for these areas requires better models.

Keywords Electricity demand · Working-class districts · Cities · Senegal

1 Introduction

In 2018, the electricity use in the residential sector represented 27% of total electricity consumption worldwide (IEA, 2020). Households use energy for various purposes, including space cooling, water heating, cooking, and lighting. In developing countries, households mainly use energy for cooking and heating services; solid biofuels and liquefied petroleum gas being the predominant fuels. Other services, including lighting, space cooling, and other plugging appliances, use electricity (Poznaka et al., 2015). The literature mentions various factors that influence electricity consumption behaviour in the residential sector. These factors

---

S. Sow (✉) · C. M. Kebe · A. Ndiaye
LE3PI–UCAD Ecole Superieure Polytechnique, Dakar, Senegal
e-mail: Sow.salif@esp.sn; ab.ndiaye@univ-zig.sn

A. Fall, R. Haas (eds.), *Sustainable Energy Access for Communities*,
https://doi.org/10.1007/978-3-030-68410-5_12

include local weather, dwellings type, number of inhabitants, and income. In Senegal, the residential sector consumed 11.7% of the total final energy consumption of fossil fuels and 25% of the electricity produced in the country in 2018. The share of electricity demand by the residential sector is projected to increase by a factor of 3 by 2040 (IEA, 2019).

Seasons and geographic location are determinants of energy consumption. Ndiaye et al. (2017) found that electricity consumption in Dakar and Niamey is relatively low during the rainy season (June–September) compared to other periods of the year. Consumption peaks are reached during the period from March to May when temperatures are high, followed by a decrease in consumption during the rainy season, which leads to another consumption peak arising in October that is the hottest month of the year in the Sahel region, with high demand for space cooling. Conversely, space heating represents the biggest share of energy end-use, with more than 50% of domestic energy consumption in OECD countries. An analysis of electricity consumption and households' behaviour in Swedish rented apartments showed that income is the main determinant of electricity consumption (Vassileva et al., 2012).

Besides economic conditions, the geographical location associated with households also affects their electricity demand. In China, Xinye Zheng et al. (2014) noted that urban dwellings consume more energy than dwellings located in rural areas. Daioglou et al. (2012) investigated energy consumption in both urban and rural areas of developing countries. They found that in both areas, cooking is predominant in end-use services, but other services that use electricity, such as space cooling and plugging appliances, have increasing shares.

Estiri (2014) identified four parameters that affect electricity consumption in the residential sector: weather, location, income, and plugging appliances. This study aims to investigate the weight of these parameters on electricity consumption in the residential sector, taking as examples the districts of Grand-Yoff and Grand-Dakar, which are two working-class areas located in the city of Dakar, Senegal.

2 Methodological Approach

2.1 Data Collection

The study methodology consists of an analysis of data collected within the framework of a study conducted by the Senegalese Agency in charge of energy savings (AEME). In 2016, AEME commissioned a survey in Grand-Dakar and Grand-Yoff in partnership with the German cooperation (GIZ). The survey sample was 120 dwellings located in these districts. The survey collected both quantitative and qualitative data in order to understand households' behaviour related to electricity consumption.

The qualitative data of the survey were collected during interviews with heads of households and include information on the demographic profile of their residence, characteristics of plugging appliances in the household and the level of knowledge on best practices in improving energy efficiency.

The quantitative data of the survey were collected during campaigns to measure the electricity consumption of the three main plugging appliances: fridge, freezer, and television (TV), and to measure the household consumption profile per week. In addition, data such as the number of plugging appliances per category were collected. The survey also collected information on the electricity consumption of these dwellings recorded by the utility (Senelec) over a period of one year.

2.2 The Study Area

Grand-Yoff and Grand-Dakar are two of the 19 districts of the city of Dakar, Senegal. Figure 12.1 displays the position of the districts on the Dakar map.

Grand-Yoff is a working-class district located in the municipality of Parcelles Assainies. According to the 2013 census, Grand-Yoff had a population of 185,503. The population was projected to reach 200,000 by 2020 (ANSD, 2016). Grand-Dakar is part of the municipality of the same name located in south-centre of Dakar. The population of Grand-Dakar was 47,012 in 2013 and was projected to reach 55,000 by 2020 (ANSD, 2016). The survey targeted dwellings in four sub-districts in Grand-Dakar and ten sub-districts in Grand-Yoff.

2.3 Data Analysis

This study focuses on the analysis of qualitative and quantitative data related to two themes: characteristics of plugging appliances and socio-economic parameters in household energy behaviour.

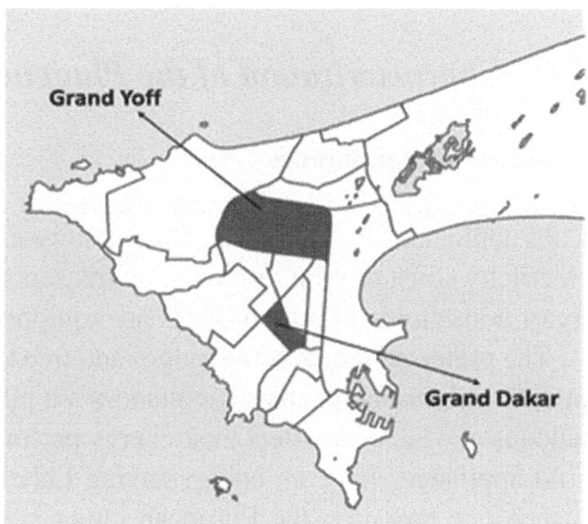

Fig. 12.1 Survey districts

2.3.1 Characteristics of Plugging Appliances

The processing of data provided information on the number of appliances in surveyed households, their rated power, and usage time per day and per period. The first part of the study consists in analyzing the energy behaviour associated with each service considering the attributes of appliances. The electricity consumed in lighting is calculated based on the number and type of light bulbs in the dwelling. The same is done for each of the other services that include space cooling (fans and air conditioners), and other services provided by plugging appliances (TVs and fridges). The objective of the analysis is to design relevant measures that improve efficiency on energy behaviour, which could target either appliances' rated power or usage per day. A similar approach has been applied in the analysis of levers of efficiency in the electricity consumption of the residential sector in Sweden (Vassileva et al., 2012 and in China (Xinye Zheng et al., 2014).

2.3.2 Socio-Economic Parameters of Energy Behaviour

The parameters classified as social and/or economic in the questionnaire collected information on the determinants of the household's energy behaviour. For instance, the level of income in the household may explain the number and rated power of appliances used in supplying the services studied. The data was first compiled in Excel, and the analysis used statistical tests to identify the distribution pattern, whether it is normal or not. Regression methods were used to study relations between the total electricity consumption per year and each of the parameters hypothesized as determinants of electricity consumption. Similar method has been used to analyze determinants of energy consumption in the United States (Estiri, 2014), and in China (Murata et al., 2008).

3 Results and Discussion

3.1 Characterization of the Plugging Appliances

3.1.1 Cold Appliances

Cold appliances include fridges that represent an average of 60% of the household electricity consumption, followed by freezers (37%), and minibars (3%). The majority of households in both districts are equipped with one cold appliance.

The preferred location for fridges and freezers is the kitchen; but we observed that in 20% of households, these appliances are placed in open areas such as hallways or balconies, which can affect their energy performance. Twenty-two per cent (22%) of cold appliances have an energy-saving Label, which is usually the energy performance rate based on the European rating system from where these appliances are

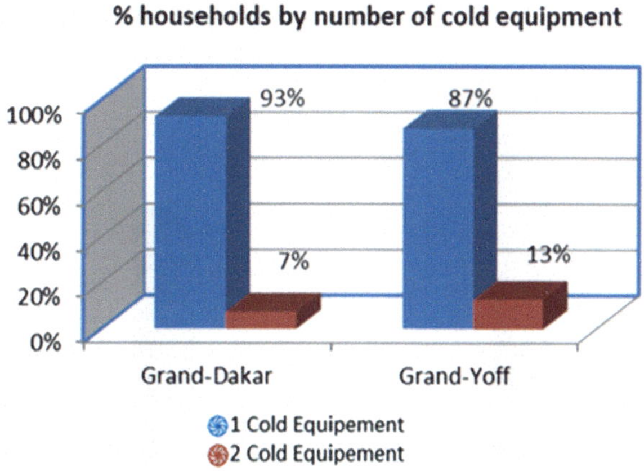

Fig. 12.2 Number of cold equipment per dwelling

exported. Thirty-six per cent (36%) of appliances have a plate indicating the climate class; out of this, 45% are of the tropical class that includes tropical T-class, subtropical ST class, and extended subtropical SN-T class. The remaining 55% are class N and SN (Fig. 12.2).

3.1.2 TV Sets

Television (TV) appliances are mainly placed in closed areas such as the living room and the bedroom. Fifty-six per cent (56%) of TVs are of the type Small and medium-sized Cathode Tube (32 inches), and 21% are of the type Plasma large screen size. For Cathode tube TV, the power measured in stabilized operating mode is 73 W; in sleep mode, this power is 7 W, making it the most energy-intensive TV on standby mode. Next is the Plasma technology TV with 4 W in standby mode and 96 W in stabilized operating mode. The LED TVs represent 14% of TV appliances inventoried during the survey, and LCD TVs represent 9% of these appliances. The LCD TV has a power measured in a stabilized operating mode equivalent to 109 W. This relatively high value is due to the brightness adjustment system of this type of screen. The majority of households surveyed have one TV set (Fig. 12.3).

3.1.3 Space Cooling Appliances

The measurement campaign took place from December to February, when temperatures are below 25 degrees Celsius in Dakar. Among the 120 dwellings surveyed, only one had a window-type air conditioner of class CV2. Fans and air conditioners were not used in any of surveyed dwellings, which prevented reliable measurements in normal operating mode. The majority of dwellings surveyed in Grand-Dakar have one fan: The situation is more contrasted in Grand-Yoff as visible in Fig. 12.4.

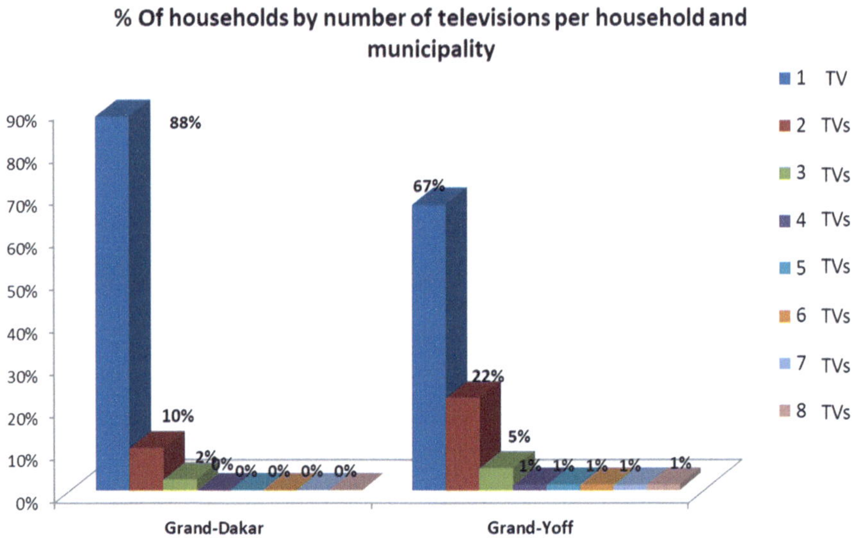

Fig. 12.3 Number of TV sets per dwelling

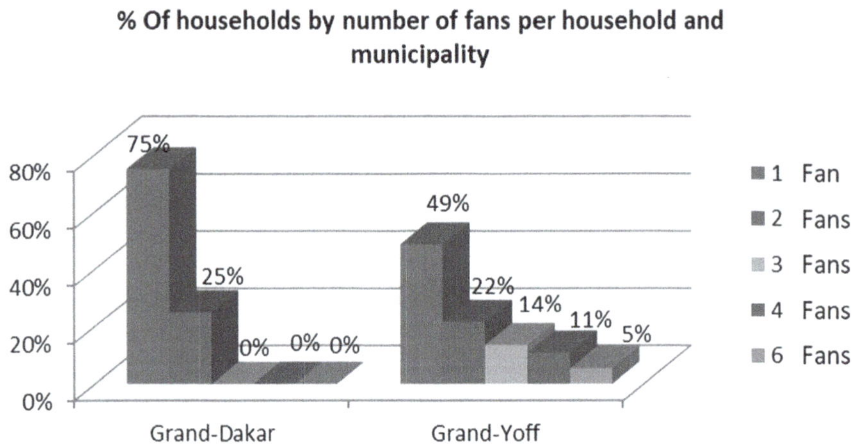

Fig. 12.4 Number of fans per dwelling

3.1.4 Lighting Appliances

The majority of dwellings surveyed have four or five lamps. Since January 2018, only energy-saving lamps that include compact fluorescent lights (CFL), halogen, and light-emitting diode (LED) lamps are authorized for sale in Senegal. The pre-legislation campaign conducted by the utility should have normally removed incandescent lamps by the time of the survey. However, 60 W and 40 W incandescent lamps were found in households, and they represent 12% of the lamps inventoried in Grand-Yoff and 25% of those inventoried in Grand-Dakar. Incandescent bulbs consume more electricity and have a relatively shorter lifespan but cost less.

3.2 *Analysis of Electricity Consumption*

3.2.1 Total Electricity Consumption Per Annum

The survey also recorded measurements of daily energy consumption from existing metres. The data collected were extrapolated in models to calculate the theoretical electricity consumption per year using the formula below:

Equation 12.1: Theoretical electricity consumption of dwellings

$$\frac{Cons\ \text{weekdays} \times 5 + Cons\ week\text{end}}{7} \times 30 \times 12$$

The theoretical electricity consumption calculated refine the data obtained from direct measurements in dwellings and project consumptions per month and per annum in the dwellings. The electricity consumption data from SENELEC invoices are labelled as the real electricity consumption of the dwelling during the same period. As shown in Fig. 12.5, the difference between real electricity consumption and theoretical estimation of electricity consumption is statistically significant.

For instance, the invoice of the dwelling coded as number 52 in the survey documents shows a real electricity consumption equivalent to 5524.13 kWh per annum, which is the highest of surveyed dwellings, but the estimation is based on

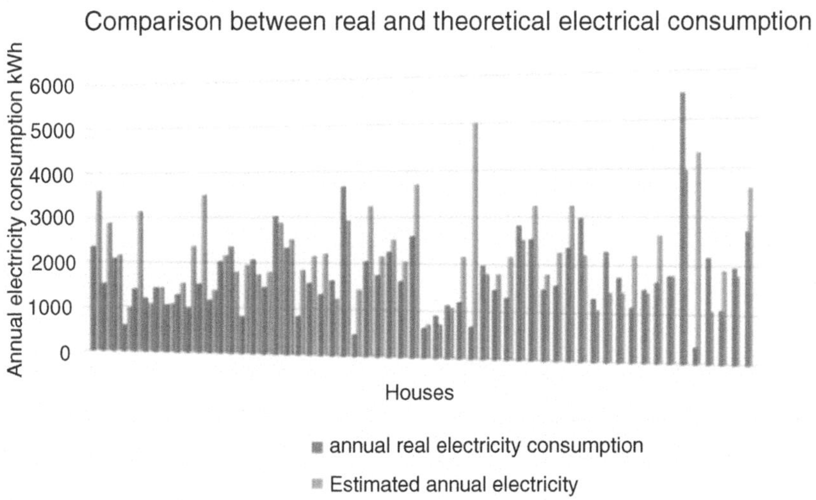

Fig. 12.5 Real and theoretical electricity consumption per annum

direct measurements return a total of 3960 kWh per annum. Considering all dwellings surveyed, the real electricity consumption represented on average 65% of the theoretical consumption. This demonstrates that planning electricity supply systems based on extrapolation of direct measurements can cause biases and results on sub-optimal systems with excess capacity, which is frequent in developing countries (Urban et al., 2007).

3.2.2 Electricity Consumption Per Capita

Per capita electricity consumption in surveyed dwellings varies between 33 kWh and 706 kWh per annum. Figure 12.6 shows variations in electricity consumption per capita in surveyed dwellings. Different parameters can explain this large difference in the electricity consumption of individuals targeted by the survey. The number of persons living in a dwelling is positively correlated with the per capita consumption ($R = 0.34$). However, the correlation is weak, suggesting at most an influence on per capita electricity consumption.

3.2.3 Impact of Social Parameters in Electricity Consumption

In this section, we look at the impact of four parameters on the real electricity consumption of surveyed dwellings. The number of residents in the dwellings, the number of residents in active employment, the level of income, and the number of appliances in the dwelling. Figure 12.7 displays the shape of plotted data.

The correlation coefficient of the parameters number of residents per dwelling and annual electricity consumption is $R = 0.34$. The correlation is weak, suggesting that the number of residents per dwelling does not significantly affect the annual electricity consumption.

The correlation coefficient of the parameters number of residents in active employment and annual electricity consumption is $R = 0.25$. The correlation is

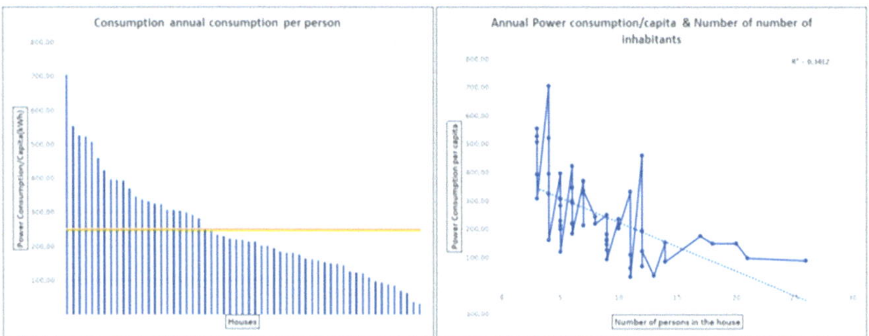

Fig. 12.6 Electricity consumption per capita

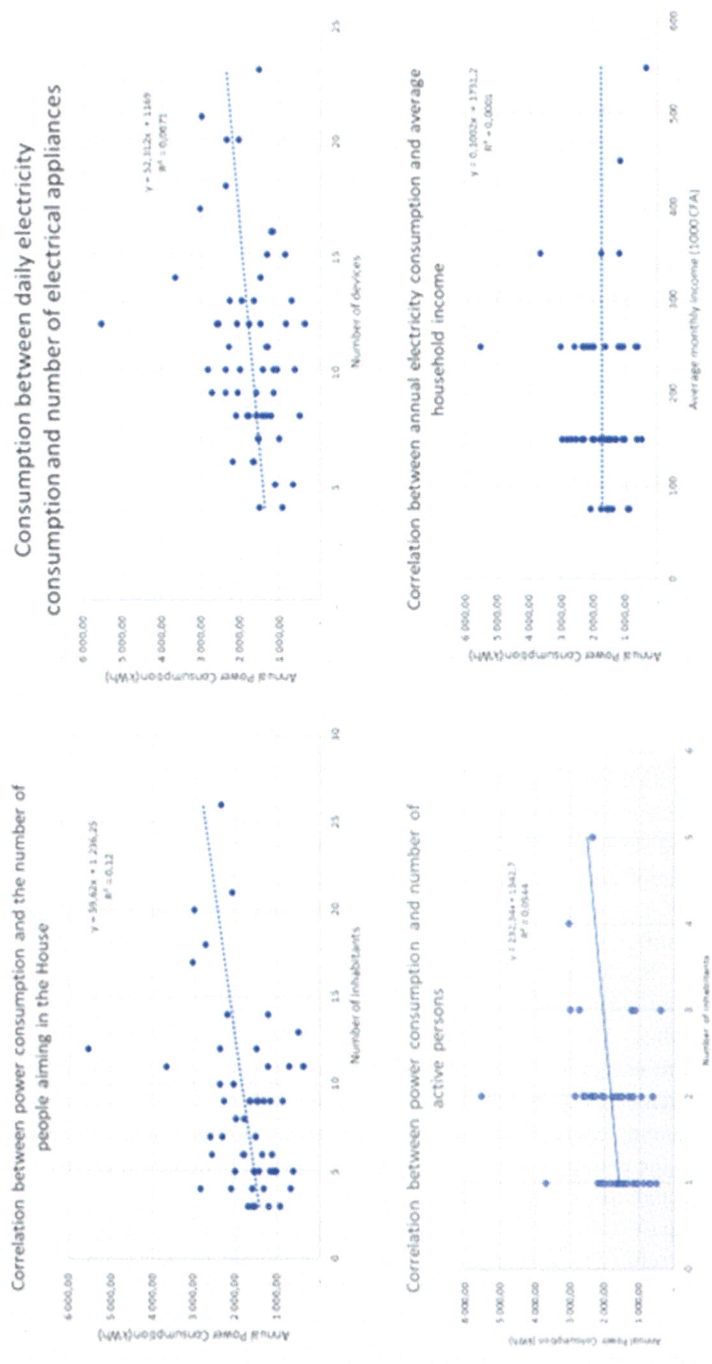

Fig. 12.7 Correlation between electricity consumption and social parameters

weak, suggesting that active employment of the dwelling's residents, and therefore absence during the daytime, does not significantly affect the annual electricity consumption.

More surprisingly, the correlation coefficient of the parameters number of plugging appliances in the dwelling and annual electricity consumption is $R = 0.25$. A higher correlation factor could have been hypothesized, with the annual electricity consumption per dwelling increasing with the number of plugging appliances in the dwelling. However, several other parameters affect the electricity consumption of these appliances, including their rated power and usage time per day.

The correlation coefficient of the parameters income of the dwelling and annual electricity consumption is the lowest of all factors at $R = 0.01$. This weak correlation suggests that electricity consumption in surveyed districts has more to do with social behaviours not identified in this study than economic purchasing power.

4 Conclusion

This chapter completed an in-depth analysis of the determinants of electricity consumption in two working-class districts of Dakar. The findings on statistical relationships between electricity consumption in these districts, with parameters that include the number of residents and the number of plugging appliances per dwelling, provide necessary information for electricity planning models. Understanding the way households consume electricity supports the design of optimal energy solutions in cities.

Findings from the study demonstrate that information on the number, rated power and usage time of plugging appliances is not enough to project the electricity demand of the dwelling. The study did not identify the parameters that have a statistically significant impact on dwellings' electricity consumption. Still, the weak correlation between this consumption and the number of plugging appliances, as well as the difference between real electricity consumption recorded by the utility and estimates based on extrapolation of periodic measurements, suggest other parameters play out in this consumption. Thus, it would be necessary to conduct advanced studies. The study provided results that are important to consider in planning future energy systems for working-class districts of cities in developing countries, which expand rapidly, driven by accelerated urbanization.

Firstly, planning energy systems at the level of city districts such as community grids through the extrapolation of data collected from a periodic survey may cause biases that would affect the size of the system.

Secondly, past invoices are not enough to project the future electricity demand of a district. Therefore, it is necessary to improve planning models with parameters different from those hypothesized in this study.

Acknowledgement The study acknowledges the support of the Agence pour l'Economie et la Maîtrise de l'Energie (AEME) for providing the data used in the analysis.

References

ANSD. (2016). *Populations des Communes de Dakar*. Agence Nationale des Statistiques et de la Demographie.

Daioglou, V., Ruijven, B. J., & Vuuren, D. P. (2012, January). Model projections for household energy use in developing countries. *Energy*, 601–615. https://doi.org/10.1016/j.energy.2011.10.044

Estiri, H. (2014, May). Building and household X-factors and energy consumption at the residential sector: A structural equation analysis of the effects of household and building characteristics on the annual energy consumption of US residential buildings. *Energy Economics, 43*, 178–184. https://doi.org/10.1016/j.eneco.2014.02.013

IEA. (2019). *Senegal Energy Outlook*. International Energy Agency. Retrieved May 14, 2020, from https://www.iea.org/articles/senegal-energy-outlook

IEA. (2020). https://www.iea.org/reports/world-energy-outlook-2020

Murata, A., Kondou, Y., Hailin, M., & Weisheng, Z. (2008, December). Electricity demand in the Chinese urban household-sector. *Applied Energy*, 1113–1125. https://doi.org/10.1016/j.apenergy.2007.05.008

Ndiaye, A., Adamou, R., Gueye, M., & Diedhiou, A. (2017, January). Global warming and heat waves in West-Africa: Impacts on electricity consumption in Dakar (Senegal) and Niamey (Niger). *International Journal of Energy and Environmental Science, 2*(1), 16–26. https://doi.org/10.11648/j.ijees.20170201.13

Poznaka, L., Laicane, I., Blumberga, D., Blumberga, A., & Rosa, M. (2015, June). Analysis of electricity user behavior: Case study based on results from extended household survey. *Energy Procedia, 72*, 79–86. https://doi.org/10.1016/j.egypro.2015.06.012

Urban, F., Benders, R., & Moll, H. C. (2007, June). Modelling energy Systems for Developing Countries. *Energy Policy, 35*(6), 3473–3482. https://doi.org/10.1016/j.enpol.2006.12.025

Vassileva, I., Wallin, F., & Dahlquist, E. (2012, February). Analytical comparison between electricity consumption and behavioral characteristics of Swedish households in rented apartments. *Applied Energy, 90*(1), 182–188. https://doi.org/10.1016/j.apenergy.2011.05.031

Xinye Zheng, C. W., Qin, P., Guo, J., Yu, Y., Song, F., & Chen, Z. (2014, December). Characteristics of residential energy consumption in China: Findings from a household survey. *Energy Policy*, 126–135. https://doi.org/10.1016/j.enpol.2014.07.016

Chapter 13
Cookinations: Mechanisms to Decouple Wood Production and Food Preparation in Sub-Urban Areas

S. Sow

Abstract In Senegal, the share of households that cook using primarily biomass fuels accounts for over 70%. Although the use of these fuels is more frequent in rural areas, there are still households in sub-urban areas that rely on charcoal. Beyond the promotion of subsidized LPG, domestic biogas and improved cookstoves are tested in some rural and sub-urban areas of the country. The results of the experiences compiled in this chapter show that these mechanisms are effective in decoupling biomass use and food preparation in sub-urban areas. Improved cookstoves can contribute to reducing biomass use by up to 45%. Domestic biogas digesters can replace biomass fuels for cooking in rural communities.

Keywords Cooking energy · Deforestation · Domestic Biogas · Improved cook stoves

1 Introduction

Cooking is a necessary part of living; every day, billions of people spend time and financial resources preparing food. Cooking activities bear different social meanings across the world. The way societies cook depends on their culture and the accessible crops. In Africa, a large proportion of the population is still dependent on small-scale agriculture for crops production (FAO, 1997). In addition, many households, especially in rural areas, rely on biomass fuels such as fuelwood, charcoal, agricultural waste and animal dung in cooking (World Bank Group, 2014). In many countries, these fuels account for about 60% of the residential sector's energy consumption (Malla & Timilsina, 2014). This reliance on biomass energy contributes significantly to greenhouse gas (GHGs) emissions and air pollution (UNICEF, 2019).

S. Sow (✉)
LE3PI–UCAD Ecole Superieure Polytechnique, Dakar, Senegal
e-mail: salif.sow01@hotmail.com

A. Fall, R. Haas (eds.), *Sustainable Energy Access for Communities*,
https://doi.org/10.1007/978-3-030-68410-5_13

In 2018, more than 2.6 billion people worldwide relied on traditional biomass fuels for cooking. Household air pollution, mostly from cooking smoke, is linked to around 2.5 million premature deaths annually]. In sub-Saharan Africa, over 80% of the population still rely on biomass fuels that include firewood and animal dung for cooking (IEA, 2020a, b, c). USAID (2016) estimates that 55.5% of Senegalese households use fuelwood for cooking, and 11% use charcoal.

Finding clean energy cooking solutions for Senegal is critical in simultaneously addressing deforestation, indoor air pollution and greenhouse gas emissions. Some initiatives have been tested over the last years to address the challenge. These initiatives include the promotion of improved cookstoves. Bensch and Peters (2019) found that improved stoves can save around 30% of firewood use in Senegal, equivalent to 27 kg of firewood saved per week and per household.

To promote a safer cooking system in Nigeria, a study piloted the use of ethanol cooking fuel in 30 households for a period of 6 months. The results show a reduction of greenhouse gas emissions estimated between 15 to 20%. In addition, to emissions reduction, the experience saved an equivalent amount of firewood and charcoal that was replaced by ethanol fuel (Ozier et al., 2018). Many other solutions aimed at reducing reliance on biomass fuel and traditional cooking devices have been tested in other regions of the world (Muok, 2018 & Diédhiou et al., 2017).

This chapter aims to document some of the experiences that effectively led to decoupling cooking energy demand and deforestation. Learning from these experiences, we propose a contribution to the development of a sustainable cooking energy agenda in sub-urban areas of Senegal.

2 Methodological Approach

The study is based on a systematic review of publications on cooking energy solutions applied in different sub-Saharan African countries. There is an extensive literature on the deployment of clean cooking solutions in different countries of the region. This study combines the lessons learned from these experiences and the results from an assessment of the study environment to propose a consistent agenda for transition to sustainable cooking energy in Senegal.

We exemplified the situation in Senegal to assess the impact of these cooking solutions on people's living standards. Most of the impact assessments found in the literature are based on stand-alone pilot projects (Ozier et al., 2018) and (Muok, 2018). The study documents the impact of improved cookstoves using the results of a GIZ project on jatropha as cooking fuel that was piloted in rural areas of Senegal (Bailis et al., 2003).

3 Discussion of Findings

3.1 Energy Demand for Cooking

In 2018, the total final energy consumption in Senegal was equivalent to 2545 thousand tons oil equivalent (IEA, 2020a, b, c). The residential sector represents about 43.5% of this energy consumption, which makes it the sector with the highest energy demand. Biomass fuel consumption was 895 thousand tons oil equivalent (ktoe). This biomass fuel includes firewood, charcoal and waste, which are mainly used for cooking and heating in small transformative activities in rural areas. The main source of cooking energy for rural households is firewood (49%), followed by charcoal (27%). These two fuels are mainly used for cooking, and a relatively small proportion is used in small transformative activities (Tchanche, 2018). In 2018, the proportion of the Senegalese population that still relied on traditional biomass fuel for cooking was estimated at 70% (IEA, 2020a, b, c). According to a study for the UN Food and Agriculture Organization (Niang, 2000), the share of energy fuels for cooking in households is as follows:

- Traditional wood: 551 kg/year per household or 52 kg/year per person
- Charcoal: 614 kg/year per household or 58 kg/year per person
- Liquefied petroleum gas (LPG): 131 kg/year per household or 14 kg/year per person

In the specific context of urban areas, a study conducted by the national agency for energy savings (AEME) estimated that 28% of households, which are mainly located in sub-urban areas, still use charcoal for cooking (AEME, 2019).

This demand for fuelwood and charcoal affects green lands in sub-urban areas and contributes to deforestation in charcoal-producing regions that are predominantly located in the southern part of the country. This charcoal production and use in households emit greenhouse gases and cause indoor pollution that primarily affects women and children.

3.2 Carbon Emissions from Cooking Energy

The greenhouse gas emissions associated with fuelwood and charcoal are presented in Table 13.1. These emissions are compared with emissions from other fossil fuels (GIZ, 2019).

Table 13.1 Estimation of GHG emission trends in Senegal

Emissions in million eq-CO_2	2005	2010	2015	2020	2025	2030
Due to fossil fuels		4.53	5.14	6.97	8.12	9.47
Due to biomass fuels		7.85	9.96	14.35	17.22	20.67
Total		12.39	15.11	21.35	25.35	30.13

Combustion of wood is often labelled as carbon neutral, considering that the carbon dioxide (CO_2) released during the process is equivalent to the CO_2 previously sequestered. Nevertheless, according to Laganière et al. smokestack (Laganière et al., 2015), CO_2 emissions during wood combustion can be 2.5 times higher than emissions from natural gas and 30% higher than emissions from coal per unit of energy produced.

Regarding indoor pollution associated with fuelwood and charcoal consumption, the World Health Organization (WHO) estimates that on average and depending on the type of device, charcoal releases the largest amount of carbon dioxide with 2559 g/kg, followed by liquefied petroleum gas (LPG). Firewood, with 1610 g/kg, is the least carbon intensive of the three fuels (WHO, 2014). However, liquefied petroleum gas contributes to efforts to combat deforestation, and therefore increases carbon sinks. The liquefied petroleum gas promotion programmes were launched back in the 1970s in Senegal. In 2018, LPG accounted for 29% of cooking fuels and is projected to reach 36% by 2030 (IEA, 2019). An increase of the LPG shares in cooking fuels was observed despite the gradual removal of direct subsidies on the fuel since 2010; the only subsidy maintained being the value-added tax (VAT) waiving (Ministere en charge des Energies, 2018).

3.3 Energy Solutions for Clean Cooking in Senegal

3.3.1 Domestic Biogas

Biogas production in rural communities is one of the answers to the problems associated with high reliance on traditional biomass, such as deforestation and land degradation. The Senegal National Biogas programme launched in 2009 capitalizes on many achievements with a large national geographic coverage. Today, many rural households have bio-digesters for the production and use of biogas for cooking and lighting purposes with cow dung as fuel. This experience had both social and environmental impacts. Biofuels reduce greenhouse gas emissions and can contribute to achieving self-sufficiency in energy, among other things.

The Senegal National Biogas project targets the installation of 10,000 households across the countrys various rural communities. The domestic bio-digesters have a capacity of 12 to 18 m3 depending on the size of the household and produce about 3.5 m3 of gas per day. In 2016, 1300 bio-digesters were installed. The objective of the second phase covering the period 2015–2030 is to build 60,664 bio-digesters. The two phases should reduce greenhouse gas emissions with saved 293,392 tonnes of CO_2-equivalent per year saved and a total of 1,975,264 tonnes CO_2-equivalent by 2030 (Fig. 13.1) (Ministere en charge des Energies, n.d.).

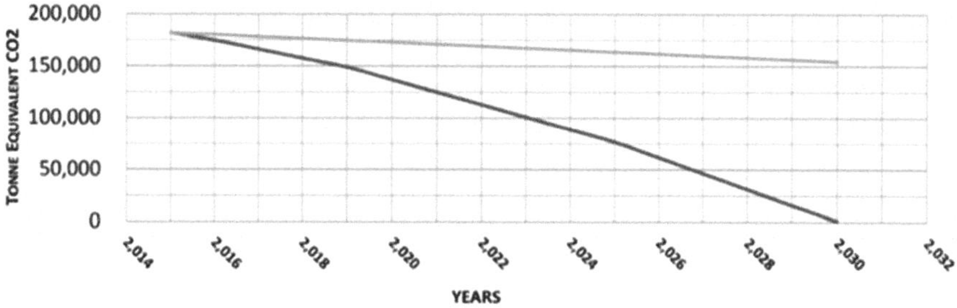

Fig. 13.1 Estimation of CO_2 emissions reduced with the Senegal National Biogas Project

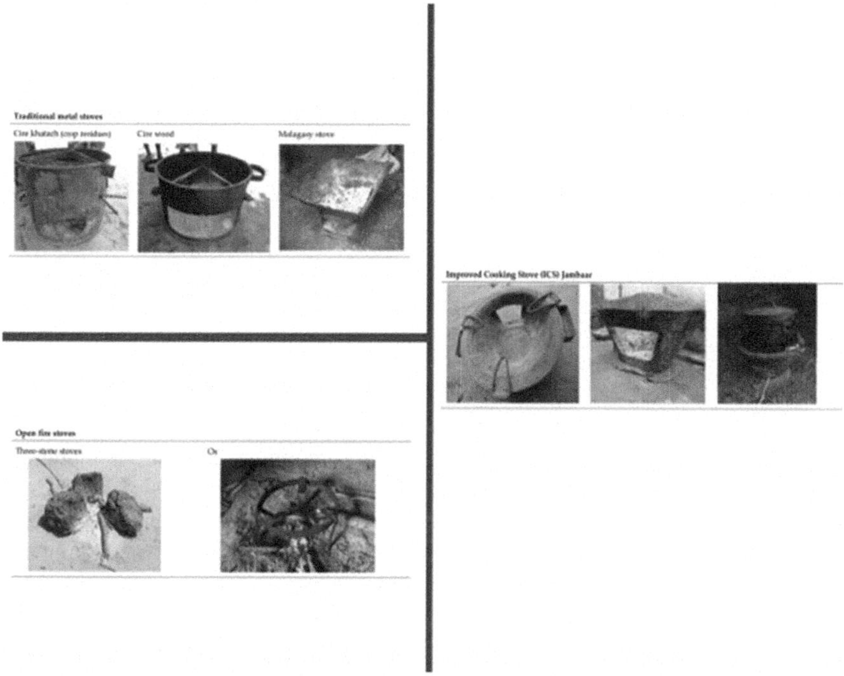

Fig. 13.2 Examples of traditional and improved cookstoves. Source: GIZ (2019)

3.3.2 Improved Cooking Stoves

The improved cooking stove is a cooking device with higher energy efficiency compared to the traditional stove, thanks to a better heat transfer to the pot by convection and radiation. The difference between the traditional cooking stoves and the improved ones is presented in Fig. 13.2 (Bensch & Peters, 2019). The promotion of improved cooking stoves aims to: (1) reduce the pressure of the demand for firewood on the forest massifs, (2) improve the quality of indoor air, (3) reduce the share of the cooking energy bill in the budgets of households, and (4) reduce the time spent in firewood collection (Institut de la Francophonie pour le Developpement Durable, 2019).

The improved cookstoves save between 30% and 45% fuel compared to the traditional fireplace and reduce smoke pollution (Putti et al., 2015). This translates into a reduction in deforestation and greenhouse gas emissions. This saving in the Senegal context means that consumption of biomass fuel could decrease by an average:

- Firewood: 200 kg/year per household
- Charcoal: 230 kg/year per household

In terms of greenhouse gas emissions (GHG) reduction using emission factors provided by the UN Food and Agriculture Organization, the equivalent reduction of GHG in Senegal is as follows:

- Firewood: 322 kg/household
- Charcoal: 588 kg/household

These results confirm that improved cooking stoves can play an instrumental role in reducing pressure on forest resources and in reducing greenhouse gas emissions in Senegal.

4 Conclusion

Cooking in Senegal is still highly dependent on traditional biomass that includes charcoal and firewood. This dependence is more important in rural communities. However, there are still households in sub-urban areas that rely on wood and charcoal to prepare their daily meals. The widespread dissemination of domestic bio-digesters and improved cooking stoves in a market-based approach targeting the overall value chain, from technology design to sale, is an opportunity to create jobs and preserve ecosystems and contributes to improve the health of women and children. These solutions also contribute to curb the degradation of forest resources and enhancing their carbon capture and sequestration potential. The limited access to clean cooking energy in Senegal (about 30% in 2018) can improve significantly with the upscale of programmes for dissemination of improved cookstoves in sub-urban areas and bio-digesters in rural areas. By 2030, domestic biogas can reduce emissions by 1,975,264 tonnes eq-CO_2. The improved cooking stoves can reduce wood consumption and CO_2 emissions by 30 to 45%. Both results contribute to Senegal' intended contribution to climate change mitigation.

References

AEME. (2019). *Service aux usagers*. Dakar, Senegal, Récupéré sur https://aemesn.com/services-aux-usagers/

Bailis, R., Pennise, D., Ezzati, M., Kammen, D. M., & Kituyi, E. (2003). *Impacts of greenhouse gas and particulate emissions from Woodfuel production and end-use in sub-Saharan Africa*. Nairobi. Retrieved from https://ecommons.aku.edu/eastafrica_eai/44/

Bensch, G., & Peters, J. (2019). One-off subsidies and long-run adoption: Experimental evidence on improved cooking stoves in Senegal. *American Journal of Agricultural Economics, 102*(1), 72–90. https://doi.org/10.1093/ajae/aaz023

Diédhiou, I., Diallo, D., Mbengue, A. A., Hernandez, R. R., Bayala, R., Diémé, R., ... Sène, A. (2017). Allometric equations and carbon stocks in tree biomass of Jatropha curcas L. in Senegal's Peanut Basin. *Global Ecology and Conservation, 9*, 61–69. https://doi.org/10.1016/j.gecco.2016.11.007

FAO. (1997). *Agriculture food and nutrition for Africa - a resource book for teachers of agriculture*. UN Food and Agriculture Organization. Retrieved from http://www.fao.org/3/W0078e/w0078e05.htm

GIZ. (2019). *Promotion of climate-friendly cooking: Kenya and Senegal*. Retrieved from www.giz.de: https://www.giz.de/en/worldwide/85477.html.

IEA. (2019). *Senegal Energy Outlook*. International Energy Agency. Retrieved November 2020, from https://www.iea.org/articles/senegal-energy-outlook

IEA. (2020a). *Key energy statistics 2018: Senegal*. International Energy Agency. Retrieved December 2020, from https://www.iea.org/countries/senegal

IEA. (2020b). *SDG7: Data and projections*. International Energy Agency. Retrieved from https://www.iea.org/reports/sdg7-data-and-projections

IEA. (2020c). *Senegal: Key indicators and policy initiatives*. International Energy Agency. Retrieved December 1, 2020, from https://www.iea.org/articles/senegal-energy-outlook

Institut de la Francophonie pour le Developpement Durable. (2019). *Les foyers ameliores*. Dakar, Senegal. https://www.ifdd.francophonie.org/wp-content/uploads/2019/12/519_Fi_foyers_ameliores_7-1.pdf

Laganière, J., Paré, D., Thiffault, E., & Bernier, P. Y. (2015). Range and uncertainties in estimating delays in greenhouse gas mitigation potential of forest bioenergy sourced from Canadian forests. *GCB-Bioenergy*. https://doi.org/10.1111/gcbb.12327

Malla, S., & Timilsina, G. R. (2014). *Household cooking fuel choice and adoption of improved Cookstoves in developing countries: A review*. World Bank Group. Retrieved from https://openknowledge.worldbank.org/handle/10986/18775

Ministere en charge des Energies. (2018). *Rapport SIE 2016*. Ministere en charge des Energies.

Ministere en charge des Energies. (n.d.). *Cadre légal et réglementaire de la bioénergie: Experience du Programme National de Biogaz Domestique du Senegal*. Accra, Ghana. Consulté le October 14, 2020, sur http://www.globalbioenergy.org/fileadmin/user_upload/gbep/docs/2017_events/5th_Bioenergy_Week_Accra_22-24_June_2017/Session_1_Birame_Faye_Senegal_legal_and_regulatory_experience.pdf

Muok, B. O. (2018). Developing innovative and sustainable market-based Systems for Enhancing Diffusion of clean cooking Solutionsin Kenya. *International Journal of Sciences, 7*(6). https://doi.org/10.18483/ijSci.1716

Niang, A. (2000). *Rapport d'étude sur les données du bois-énergie au Sénégal*. UN Food and Agriculture Organization. Récupéré sur. http://www.fao.org/3/X6799F/X6799F00.htm

Ozier, A., Charron, D., Chung, S., Sarma, V., Dutta, A., Jagoe, K., ... Olopadea, C. O. (2018). Building a consumer market for ethanol-methanol cooking fuel in Lagos, Nigeria. Energy for Sustainable Development, 46, 65–70. doi:https://doi.org/10.1016/j.esd.2018.06.007.

Putti, V. R., Tsan, M., Mehta, S., & Kammila, S. (2015). *The state of the global clean and improved cooking sector*. Washington D.C. Retrieved from https://openknowledge.worldbank.org/handle/10986/21878

Tchanche, B. (2018, January). Analyse du système énergétique du Sénégal. *Revue des Energies Renouvelables*, 21(1), 73–88. Récupéré sur https://www.cder.dz/download/Art21-1_9.pdf
UNICEF. (2019). *Silent suffocation in Africa: Air pollution is a growing menace*. United Nations International Children's Emergency Fund. Retrieved from https://www.unicef.org/reports/silent-suffocation-in-africa-air-pollution-2019
USAID. (2016). *Greenhouse gas emissions factsheet: Senegal*. U.S. Agency for International Development.
WHO. (2014). *WHO guidelines for indoor air quality: Household fuel combustion*. WHO. Retrieved October 15, 2020, from https://www.who.int/airpollution/guidelines/household-fuel-combustion/en/
World Bank Group. (2014). *Clean and improved cooking in sub-Saharan Africa: A landscape report*. World Bank Group. Retrieved from https://openknowledge.worldbank.org/handle/10986/22521

Chapter 14
Citizen Awareness of the Social Dimension of Energy: Lessons from a Survey in Dakar

Cheikh Faye and M. Wade

Abstract The 2018 uprising of "Gilets jaunes" in France gave prominence to the question of whether to fund the energy transition or lift up the buying power of populations. This topical concern raises older and yet unsolved questions, which are how much does cost the energy transition and who should bear the bill. The chapter explores emerging issues in the challenge of transition to energy sustainability in communities and focuses on citizens' perception of the social dimension of energy, besides the economic and environmental dimensions. The study findings from a survey conducted in Dakar show that citizens' opinions on energy are shared between necessity—for living—and concern about its pollution. Our findings support the design of a localized communication strategy to support the energy agenda for transition to sustainability.

Keywords Energy survey · Energy behaviour · Social sustainability

1 Introduction

Energy is the primary contributor to greenhouse gas (GHG) emissions and contributes significantly to global warming and loss of biodiversity. Access to large and cheap energy resources contributed to the most important human innovation of the last decades, including in industry an in communication. Humankind is at a pivotal time to address constant energy challenges related to both positive and negative externalities of use. In this chapter, we explore the social dimension of these challenges at the

C. Faye (✉)
Ministry of Labour, Business Integration Programme, Dakar, Senegal
e-mail: cheikhfaye990@gmail.com

M. Wade
Universite Cheikh Anta Diop Dakar, Dakar, Senegal

Ecole Polytechnique Thies, Thies, Senegal
e-mail: mwade@ept.sn

A. Fall, R. Haas (eds.), *Sustainable Energy Access for Communities*,
https://doi.org/10.1007/978-3-030-68410-5_14

community level. Our ambition is to capture how citizens' perception of challenges related to externalities of energy use could inform strategies of transition to energy sustainability in local communities. The concept of local action on energy is linked to that of a territory where humans and infrastructure power dynamics towards a form of development that could be more sustainable with the exploitation of locally available renewable energy resources. To capture this perception, we prepared and conducted a survey in two districts of Dakar between April and May 2019.

The literature on energy sustainability rarely addresses the social dimension of energy through the prism of individual citizens and their opinions on attributes associated to energy as a resource, as a good, and as a fuel.

Bally (2015) investigates the concept of citizens' initiative in energy and highlights the gap between public institutions and the individual actors. The bottom-up method he proposed starts with a compilation of the initiatives that individual actors can embody in the sector as examples of relevant actions for public institutions. According to Buclet (2011), it is important to measure and analyze the results obtained by these local initiatives so that they are useful for the actors involved and ultimately contribute to their improvement. The evaluation must, therefore, support or even legitimize these emerging processes in order to make them accessible to public action.

Ayyoob (2020) explores indicators such as economy, governance, mobility, environment, population, habitat, and data in the development of smart cities in order to improve accuracy using information technologies. He discussed the indicators and proposed areas for improvement regarding the methodology for evaluating these indicators. The methodology should include a broad research strategy with a wide range of evaluation schemes.

These studies focus, respectively, on citizens' initiative and on the potential for improving the methodology used to collect data on indicators related to these initiatives. They did question neither the individual actors mentioned in the studies nor their perception of these indicators and did not document the background of citizens' commitment to initiatives in the energy sector. Those are the questions that this study aims to address. The chapter explores citizens' perception in three dimensions of energy use: energy efficiency in buildings, recycling of waste to energy, and non-quantitative parameters that citizens' associate with energy.

2 Methodological Approach

The study method is a survey conducted in 245 residence buildings located in two districts of Dakar in Senegal, namely the districts of Amitié and Fann-Point E. The survey was conducted within the framework of the project Sustainable Energy Access for Sustainable Cities (SEA4cities). The five SEA4cities scholars conducted the study with the support of local populations in target districts. The survey took place from April to May (Amitie and Fann-Point E). The survey sample size is based on population figures in the target districts provided by the National Statistics and Demography Agency (ANSD), and hypothesizing a confidence level of 95%.

The method is complementary to that of Bally (2015), who proposes a number of initiatives that could carry individual citizens as a source of information for decisions by public institutions to improve energy sustainability. Our survey method, which targets citizens brings in the process of initiatives compilation, the social context. In addition, our approach addresses potential biases that could arise from disconnection between models from literature and information from the field that connects energy behaviour with social parameters such as age, gender, level of education, and the communities' specificities.

The method is different from that of Ayyoob (2020), who focuses on a broad research strategy with diagrams to assess the intelligence of cities as a source of information on the identification of relevant indicators; the periodic assessment of these indicators being the way to inform on their effectiveness in guiding cities towards a smarter, more sustainable and climate-resilient future.

Our method that consisted in questioning citizens directly on their energy situation, daily behaviour and opinions aims to reposition the social dimension of energy within society and to answer the question of energy meaning for energy users. Our approach consisted in relaying opinions of these users before analyzing what these opinions could mean. The collection and analysis of data form citizens involve three phases: pre-survey, survey, and post-survey.

2.1 The Pre-Investigation Phase

During this phase, we computed the study sample using data from the National Statistics and Demography Agency (ANSD), municipalities and other national agencies that store data related to buildings' functions. The questionnaire was also prepared during this phase. The questionnaire requested data on the following parameters:

- Technical parameters: energy appliances (type and number), daily usage (hours of use per day), consumption fuels (e.g. electricity, LPG, charcoal), period of usage (e.g. ventilation appliances and comfort temperature), and waste quantities.
- Socio-economic parameters: energy budget, household income, perception of the social and environmental attributes of fuels.

2.2 The Investigation Phase

The main issue during this phase was to minimize potential biases on data collected. Therefore, it was necessary to ensure the survey coincided with the special periods of Ramadan and Eid to capture their impact on energy consumption. The questionnaires with answers were scanned on a daily basis and stored in the SEA4cities drive.

2.3 Post-Survey Phase

Following the collection phase, the data were compiled in an Excel format matrix as follows:

- In rows: the code identifying the building: RCD_res_number for residence buildings.
- In columns: the quantitative and non-quantitative data.

The quantitative data group includes building parameters and the indicators associated with energy services, such as type of lamps and hours of use per day for the lighting service. Data on four services were reported: lighting, ventilation, cooking, and waste management. Another column compiles energy data relating to other uses such as modems, TV sets, and printers.

The non-quantitative data group includes attributes associated with fuels consumed in the building. Three fuels were proposed in the questionnaire: electricity, liquefied petroleum gas (LPG) and charcoal. The social and environmental attributes were also pre-defined, and respondents were requested to choose the first three they associate with each of these fuels.

3 Results and Discussions

The analysis of the matrix compiling the data collected returned the following theme-based results.

3.1 Energy Efficiency in Buildings

For the first theme that looked into the potential of energy efficiency in buildings, we analyzed three (3) indicators:

- Energy behaviour with respect to lighting
- Energy behaviour with respect to ventilation
- Energy behaviour with respect to other appliances recorded in buildings

We computed the average energy consumption of the 245 buildings surveyed, equivalent to 3620.27 kWh. The ventilation consumption share is 1311.22 kWh, while lighting consumed 602.75 kWh, and other appliances 1706.33 kWh. Thus, the share of energy consumption for ventilation in surveyed buildings is equivalent to 36.21%, and that of lighting is 16.6%.

3.2 *Recycling Waste to Energy*

With respect to the second theme that looked into the potential of waste conversion to energy, we analyzed three (3) indicators:

- The waste production per building (in kg /year).
- The energy potential of waste recycled by gasification (in kWh/year).
- The energy potential of waste recycled by incineration (in kWh/year).

3.2.1 Potential Waste per Household (Kg/Year)

The first remark after processing data collected to feed these indicators is that the quantity of waste produced in the building is poorly correlated with the size of the building. The correlation factor is 0.0095 and suggests that the production of waste is not significantly correlated with the size of the building. Secondly, the data show significant discrepancies in terms of the quantities of waste produced; After extrapolation of daily quantities, the building with the highest quantities can produce 4745 kg per year, while the buildings with the lowest quantities have a potential of 380 kg per year. The average waste quantities produced is 1448 kg per year and per building.

3.2.2 Potential of Energy Recovery from Gasification

The potential of energy recovered from the waste produced by the 245 buildings surveyed is computed considering the average share of biodegradable components on the total quantities assumed at 13%, considering UCG (2016) figures. The energy potential from gasification is 1154 kWh per year.

3.2.3 Potential of Energy Recovery from Incineration

The potential of energy recovered by incineration is established considering the average share of the waste components that can be incinerated, meaning plastics, papers, and wood clogs assumed at 46.7% of the total quantities considering UCG (2016) figures. The average potential is 4145 kWh per year.

The analysis of district data shows a lower proportion of biodegradable waste in Fann-Point E and Amitié compared to proportions observed in the district of Diamniadio that was surveyed before. Therefore, incineration is more suitable for these districts that also consume on average more electricity than the district of Diamniadio.

3.3 Non-Quantitative Parameters of Energy Consumption

With regard to the third theme that looked into non-quantitative parameters of energy consumption, we analyzed two (2) indicators:

- Fuel classification by attribute (perception of surveyed consumers)
- Fuel classification by pollutant (perception of surveyed consumers)

3.4 Fuel Classification by Attribute

The compilation of attributes the respondents associate with LPG, charcoal, and electricity, using the pivot table, returns Figs. 14.1, 14.2, and 14.3.

3.4.1 LPG

About one-third (31.02%) of households consider LPG as necessary. LPG is used for cooking in households. In the 1970s, the government of Senegal decided to promote LPG for cooking through an intensive subsidies policy that makes the fuel as accessible as charcoal, especially in urban areas, in order to address the overexploitation of biomass resources for charcoal production. We also noticed that over 29% of respondents do not have an opinion on this fuel. They use it because it is available in the market. Only 2% associate the fuel with reliable, which might be due to frequent shortages during the period 2009–2011. Two respondents consider competing alternatives are available in the market, the main alternative being charcoal.

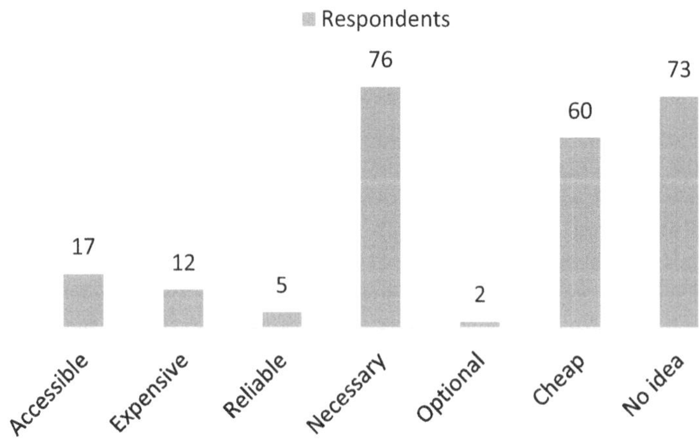

Fig. 14.1 Attributes associated with LPG

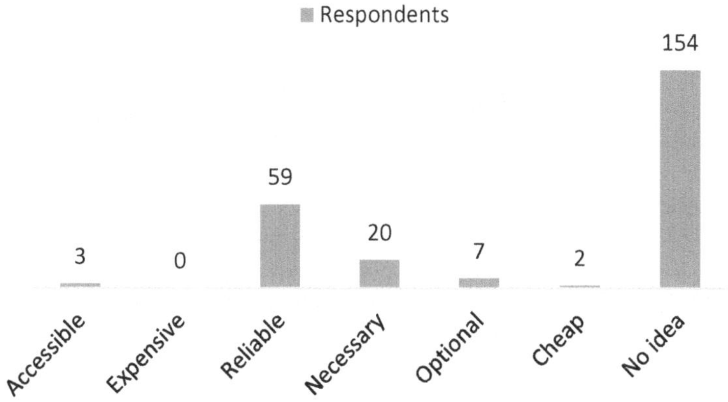

Fig. 14.2 Attributes associated with charcoal

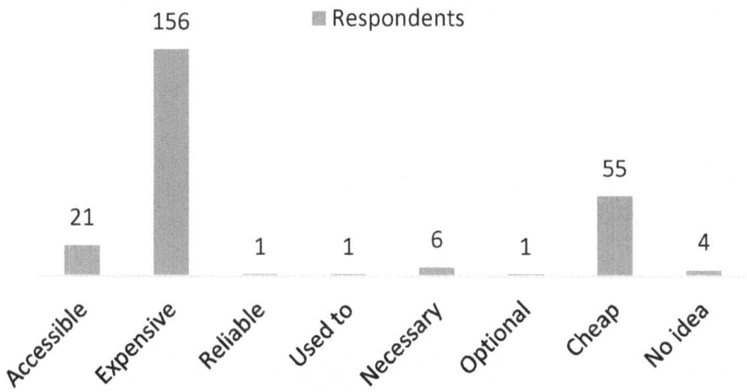

Fig. 14.3 Attributes associated with electricity

3.4.2 Charcoal

The majority of respondents (62.9%) do not have an opinion about charcoal. This relatively high figure may relate to the difficulty to access charcoal in the market nowadays. LPG was promoted in conjunction with a repressive policy on charcoal production and sale. In Dakar, charcoal is sold in small quantities in shops and the price is often prohibitive for large use. Still, 24% of respondents perceive the fuel as reliable. The main argument that comes with the answer is even though charcoal is expensive, it has been an alternative during periods of LPG shortages. Less than 1% of respondents associates charcoal with "Accessible" and "Affordable."

3.4.3 Electricity

The majority of respondents (63.7%) associate electricity with "Expensive." The electricity tariff for residential buildings in Dakar during the period 2016–2019 was EUR 18 cents per kWh. This is less than the tariff in Vienna, which was about EUR 22 cents for residential buildings, but the level of revenues is much lower in Dakar. We also noticed electricity is the fuel that aroused more enthusiasm from respondents. Only1.6% said they did not have an opinion about the fuel. About 9% associate it with "Accessible." This proportion may be due to the fact all buildings surveyed were connected to the grid and this connection was not always established by respondents to the questionnaire.

Four (4) out of 245 surveyed have decentralized solar rooftop systems operational in their buildings. Respondents were requested to note the four stages relating to the operation of these decentralized production systems, namely: acquisition, installation, operation, and maintenance. One (1) respondent found the acquisition of the system difficult, but the majority did not make comments on the acquisition. Three out of four respondents did not comment on the maintenance of their decentralized system.

3.5 Fuel Classification by Pollutant

Answering the question on their perception of the energy sector's ecological footprint, the respondents associated their energy consumption with a list of pre-defined pollutants proposed in the questionnaire. The classification was in a ranking order from the first selection (more concerning) to the third selection (less concerning). The compilation of pollutants that the respondents associate with their energy

Table 14.1 Pollutants ranked first

Pollutants	
GHG	121
Smoke	19
Odour	3
No idea	102
Total	245

Table 14.2 Pollutants ranked second

Pollutants	
GHG	0
Smoke	53
Odour	3
No idea	189
Total	**245**

Table 14.3 Pollutants ranked third

Pollutants	
GHG	0
Smoke	0
Odor	52
No idea	193
Total	245

consumption in ranking order, using the pivot table, returns Tables 14.1, 14.2, and 14.3.

Majority of respondents (49.4%) associate their energy footprint with greenhouse gas (GHG) emissions. This perception might be related to the connection the media make between energy, global warming and greenhouse gas emissions. Dakar is highly impacted by sea-level rise associated with global warming. The city is located at about 22 m above sea level and has lost many coastal areas over the last decades, which has been largely reported in media.

The majority of respondents (77.1%) did not make a connection at this level. 37.1% of respondents who selected a pollutant in the first level of ranking selected smoke as the second pollutant associated with their energy consumption. The smoke in question is that of power units that are visible in the city entrance (Rufisque), and, to a lesser extent, the smoke generated by charcoal use in buildings that reported use of this fuel.

The majority of respondents (78.8%) did not make a connection at this level. 92.8% of respondents who selected pollutants in the two previous levels of ranking selected odour as the third pollutant associated with their energy consumption. The odour, in question, is that of LPG bottles and charcoal fuels during operation.

These findings based on the analysis of the data collected during the survey demonstrate, first and foremost, ordinary citizens' interest in energy-related issues. This interest was already observed during the data collection process where interviewees raised questions about their energy situations, the possibility to access alternatives to electricity supply from the grid, the status of renewable energy resources, tariffs, etc.

Citizens are aware of the social dimension of energy but usually fail to see it in a larger perspective. For instance, LPG shortages during the period 2009–211, which explains why this fuel is still perceived as unreliable, is largely due to the global energy crisis of 2009, which also affected the charcoal cost. However, citizens defended the narrative of a perverse policy that preserves trees to the detriment of humans' living standards. Surprisingly, the same citizens' voice concerns about the relations between their electricity use, from the utility, and the sea-level rise due to global warming. These observations suggest the social and environmental dimensions of transition to energy sustainability are context specific. Dakar has no large green lands and is, therefore, less affected by the consequences of deforestation. However, as a peninsula, it is sensitive to sea-level rise. Therefore, Communication to rally the necessary support of communities to energy transition agendas should have a localized approach. In fact, the energy transition agenda should be more context specific, and its promotion should link its measures with local concerns.

Surveyed citizens are sensitive to greenhouse gas emissions from power plants. A transitional regime that would reduce these emissions could probably spur their enthusiasm. Distributed renewable energy systems offer this alternative to electricity from large power plants connected to the grid. The solar and wind energy resources are readily available in the city, and the regulatory framework supports this alternative. However, despite the regulatory and fiscal incentives enforced since 2011, these systems' dissemination is still marginal, as observed during the survey, with only four buildings out of 245 equipped with decentralized renewable energy systems. Two factors may explain this gap: the perception of inaccessibility and a lack of information on real figures. Energy is still perceived in its technical dimension of wires and connections that is only accessible to professionals. In this perception, the professionals who manage the grid appear as the best alternative, despite the negative attributes associated with the product of their management such as expensive, not reliable, polluting, etc. The survey reveals a form of fatalism associated with the energy situation that inhibits the citizens' willingness to look for alternatives, including decentralized renewable energy alternatives. This is a type of crowd effect where citizens do not necessarily feel comfortable with the bigger number, but value the absence of responsibility, here in the acquisition and maintenance of decentralized energy systems. This value is due to a lack of information on parameters such as technical requirements, costs, and payback periods.

Energy cost is another parameter of interest for citizens when considering the electricity sector. Distributed renewable energy generation also contributes to lowering electricity costs, but savings only appear in the long-run. The value time of money is another issue to address in local energy agendas. Communication with citizens could, as an example, show the comparison between investment figures in T and figures of cost savings in T + 25. Beyond the communication of figures, there should be incentive mechanisms that target the financial flows between the periods of investment and cost savings by addressing the concepts of risks and solvency, which could spur on contagion effects that are specific to crowd behaviours.

4 Conclusion

The survey confirmed the citizens' interest in the social dimensions of the energy sector. The sector does not traditionally drive the same passions in government's communication as water or education, especially in regions that are familiar with energy access and supply issues. However, since the 2009 energy crisis, issues in the sector raise concerns among citizens. In Dakar, the security of supply of the three fuels analyzed in this study is a major concern. If insecurity in charcoal supply is caused by national policies that limit its production, insecurity in the supply of electricity and LPG are to associate with the global context partially. This global context is another concern for citizens in relation to climate variability and its consequences on the city environment. Dakar has been highly affected by sea-level rises, and our survey showed that the relationships scientists build between energy production and climate change, which is relayed by media, have reached the citizens. However, a question remains on how the perception of a costly and pollutant energy supply could translate into investments in sustainable energy systems.

We propose the social buy-in mechanism to engage citizens in driving the energy transition to sustainability. Buy-in by large groups of people has been discussed in numerous psychological publications, and, in this analysis, we propose to use this psychological dimension as the basis for communication on an energy agenda, reinforced with economic incentives, which would motivate citizens to adopt new energy behaviours. The social environment of cities is a dense network of relationships that is a breeding ground for crowd effects. Policies that promote the transition to energy sustainability could lever this effect by listening directly from citizens and by mainstreaming the specificities of local contexts in national agendas. This study was limited to the energy behaviour of citizens. We tried to understand the perception of energy from the consumers' perspectives through the analysis of their unfiltered answers. It is undisputable that this perception is influenced by economic and technical factors that we did not look into details.

References

Ayyoob, S. (2020, April). A typology of smart city assessment tools and indicator sets. *Sustainable Cities and Society, 53*, 105–364. https://doi.org/10.1016/j.scs.2019.101936

Bally, F. (2015). Vers unte transition energetique citoyenne: la reappropriation des questions environnementales par les acteurs ordinaires. *Rives Mediteraneennes*, 67–79. https://doi.org/10.4000/rives.4940

Buclet, N. (2011). Territoire, innovation et développement durable : l'émergence d'un nouveau régime conventionnel ? Dans A. Colin, *Revue d'Économie Régionale & Urbaine* (pp. 911–940). Paris: revues.armand-colin.com. https://doi.org/10.3917/reru.115.0911.

UCG. (2016). *Programme de gestion des dechets solides urbains de la region de Dakar*. Unite de Coordination et de Gestion des dechets.

Chapter 15
Energy in Development Objectives: How the Energy Ecological Footprint Affects Development Indicators?

W. M. Nkounga

Abstract Access to affordable energy services remains a priority for eradicating poverty in developing countries. Energy services from conventional resources are necessary to power economic growth. However, they have a significant ecological footprint. In this study, we assess the impact of greenhouse gases (GHG) emitted by energy systems on some of the Sustainable Development Goals (SDG). The study explores the relationships between the carbon intensity of the energy sector, the energy intensity of the economy and the carbon intensity of the economic system. In a sample of African countries, we found a positive correlation between energy use per capita and greenhouse gas emissions per capita and per unit GDP (carbon intensity of the economy). However, the correlation is less conclusive between energy use per capita and GHG emissions and between energy use per capita and energy use per unit GDP (energy intensity of the economy). Our results support new perspectives on energy sustainability agendas that take into account the macroeconomic parameters of the Sustainable Development Goal number 7 (SDG-7).

Keywords Carbon intensity · Energy intensity · GDP · Greenhouse gas emissions

1 Introduction

Transition to energy sustainability is a major goal of the pots-2015 development agenda adopted by countries in the sub-Saharan Africa region. Considering their level of gross domestic product (GDP) per capita, these countries are classified at different scales of development. Still, the concept of sustainability associated with that of development requires additional indicators beyond per capita GDP. Indicators of sustainable development are largely documented in the scientific literature (Eustachio et al., 2019). These indicators include energy access, primary energy

W. M. Nkounga (✉)
Université Cheikh Anta Diop de Dakar, Dakar, Sénégal
e-mail: welly-magloire.nkounga@ucad.edu; willy-magloire.nkounga@univ-smb.fr

A. Fall, R. Haas (eds.), *Sustainable Energy Access for Communities*,
https://doi.org/10.1007/978-3-030-68410-5_15

use, and greenhouse gas (GHG) emissions. GHG emissions in the energy sector are of particular concern because energy is the primary contributor to these emissions, with a share estimated at 75% (Wang et al., 2019). GHG emissions contribute to global warming and have detrimental effects on the environment and on economic sectors such as agriculture and breeding in the sub-Saharan Africa region (Esso & Keho, 2016).

According to the International Energy Agency (2020a, b, c), the reduction of global emissions worldwide anticipated at 30.6 gigatonnes (−8% compared to 2019) is mainly due to the decline of emissions from coal (−8%), oil (−4.5%), and natural gas (−2.3%) following the lockdown measures in large economies such as China and the USA. This IEA scenario shows that, in addition to the level of demand, the primary energy mix causes different countries to have differentiated levels of contributions to greenhouse gas emissions, and therefore to mitigation efforts. In its 2018 report, the United Nations' Intergovernmental Panel on Climate Change (IPCC, 2018) estimated that limiting global warming to 1.5 °C will require increasing the share of renewables in the electricity supply mix to 97% by 2050; the International Energy Agency estimated the share of renewables in the global electricity generation at 26% in 2018 (IEA, 2020a, b, c).

Different methods have been experimented with in studying the relationships between economic growth, energy use, and greenhouse gas emissions. These methods include statistical regression models, vector errors correlation models, Granger causality, multivariate statistical methods and bounds test. Using the Granger causality and the bounds test, Zaman and Abd-el.Moemen (2017) studied the correlation between economic growth, energy consumption, and CO2 emissions in a sample of sub-Saharan Africa countries, using data series of the World Bank recorded for the period 1970–2010. Their results show that economic growth is positively correlated with CO2 emissions in Benin, Cote d'Ivoire, Ghana, Nigeria, Senegal, South Africa, and Togo. Adom et al. (2012) also used the bounds test to establish a correlation between economic growth, industrialization process and greenhouse gas emissions in Ghana, Morocco, and Senegal. Their results show that industrialization supports economic growth, which is positively correlated with greenhouse gas emissions. Eustachio et al. (2019) proposed a methodology to monitor the process of sustainable development in different countries, which is based on system theory and uses data from the World Bank data series.

However, these studies are limited by the absence of recommendations on how the correlation between economic growth, energy use, and greenhouse gas emissions could inform the planning of a tailored energy agenda specific to each country. This study aims to identifying levers of action on the energy systems of eight (8) sample countries based on the structure of their economies and levels of greenhouse gas emissions associated with different sectors, including the energy sector. The results should contribute to countries' efforts in mitigating the energy sector's ecological footprint taking into account the cross-sectoral dimension of the energy sector.

2 Methodological Approach

The study approach consists in analyzing the correlation between the energy demand per capita of a country and the indicators of economic development exemplified here with the gross domestic product (GDP) per capita and the indicators of ecological footprint exemplified with the equivalent carbon dioxide emissions (CO2e) per capita. We start by collecting the data to compute these correlation factors using the online-accessible Global Change Data Lab platform. The computation integrates the following country indicators (the year 2016):

- Primary energy use per capita in MWh
- Gross domestic product per capita in US Dollars of 2011 purchasing power parity (PPP)
- Greenhouse gas emissions in equivalent CO2 emissions per capita (CO2e)
- Energy intensity of the economy (kWh per unit GDP)
- Carbon intensity of the energy sector (kg CO2e per kWh)
- Carbon intensity of the economy (CO2e emissions per unit GDP

The study exemplifies the following countries located in the sub-Saharan Africa region: Cameroon, Gabon, Kenya, Mozambique, Nigeria, Senegal, South Africa, and Uganda, which are representative of the four regional economic communities (RECs) in sub-Saharan Africa. Inside these countries, the study also looks at the energy and GHG emissions' figures associated with the sectors of agriculture, building, electricity and heat generation, land-use change and forestry, manufacturing and construction, industry, and transport.

The study starts with a presentation of the levels of CO2 and other greenhouse gas emissions in sample countries and in key economic sectors of these countries. Secondly, the study computes the correlation factors between energy demand per capita and the equivalent carbon dioxide emissions per capita, and between GDP per capita and equivalent carbon dioxide emissions per capita for each of the sample countries. From these assessments, we derive recommendations on the selection of tailored sustainable energy agendas that integrate renewable energy resources and other efficient measures to improve the carbon intensity of the countries' economies.

3 Results

Fig. 15.1 displays the levels of carbon dioxide and greenhouse gas (GHG) emissions of sub-Saharan Africa countries in the sample.

Legend: CO2 emissions are emissions from the energy and cement industries. GHG emissions include, in addition to CO2, five gases with greenhouse effect, including methane (CH4) and nitrous oxides (NOx) emitted in the energy sector and other key economic sectors.

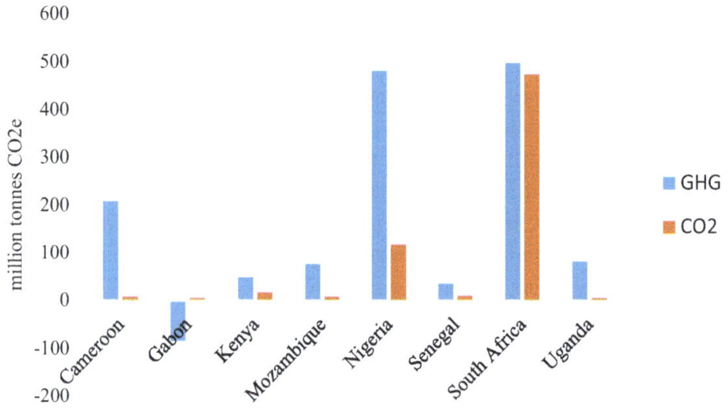

Fig. 15.1 CO2 and GHG emissions in 2016 (Global Change Data Lab, 2020a, b, c, d, e)

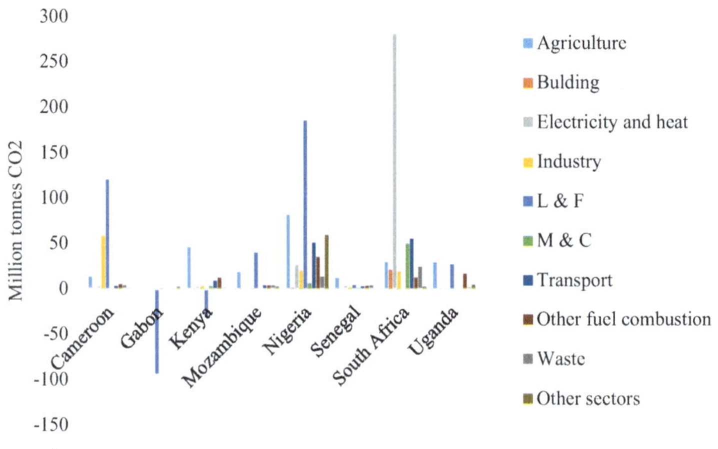

Fig. 15.2 GHG emissions per sector (Global Change Data Lab, 2020a, b, c, d, e)

Estimates of CO2 emissions range between 5.4 million tonnes in Uganda and 474.98 million tonnes in South Africa. Estimates of GHG emissions range between −84.96 million tonnes in Gabon and 497.39 million tonnes in South Africa. South Africa has the highest level of carbon dioxide emissions. However, its level of other GHG emissions is lower than in Nigeria. These other GHG emissions include emissions from land-use change and forestry and emissions from transformative industries with relatively high radiative forcing such as chlorofluorocarbons (CFCs), hydrofluorocarbons (HFCs). Their amounts are related to the number of transformative industries requiring heat or cold production. The total greenhouse gas emissions in Gabon are negative due to carbon capture by the vast forest resources of the country. GHG emissions in Kenya (47.77 million tonnes) are relatively low due to the same phenomenon of carbon capture accounted in the inventories. Figure 15.2 provides details of greenhouse gases emissions per sector.

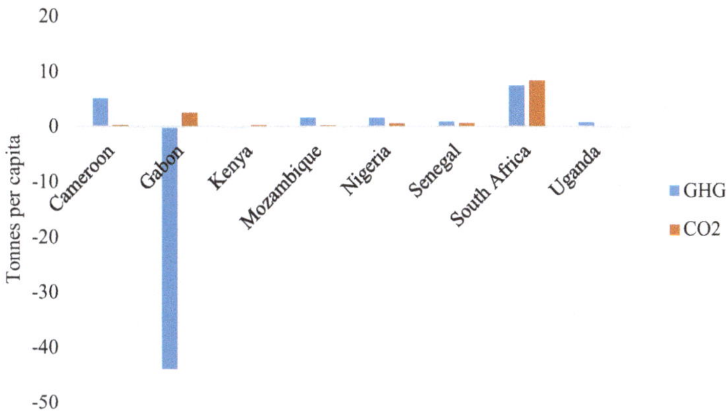

Fig. 15.3 GHG and CO2 emissions per capita (Global Change Data Lab, 2020a, b, c, d, e)

Legend: L and F stands for land-use change and forestry. M and C is the manufacturing and construction sector. Other sectors include fugitive emissions, defined as the accidental leakages of gases such as methane in the oil and gas industry.

Agriculture is the first contributor to GHG emissions in Kenya, Senegal, and Uganda, and the second contributor to emissions in Nigeria, Mozambique, and South Africa. The sector predominantly produces methane (CH4) and nitrous oxides (NOx), whose levels are higher in composting and burning of crop and grass residues.

Emissions from power plants recorded in the electricity and heat sector are relatively high in South Africa due to this country's installed capacity of 51,305 MW of which 91.2% is made of thermal units (USAID, 2020a, b).

The transport sector is an important contributor to greenhouse gas emissions in Nigeria and South Africa, where they contribute, respectively, with 50.8 and 55.4 million tonnes. The relatively high share of the sector in these countries' GHG emissions is related to the scarcity of mass transport facilities that makes private cars necessary. Traffic congestion in sub-Saharan Africa has dramatic consequences on both the environment and the macroeconomic indicators of the countries. The sector also represents a relatively high proportion of emissions in Kenya, with 18.6%. In all other countries in the sample, its contribution is below 10% of greenhouse gas emissions.

Apart from South Africa, emissions from the building sector and the manufacturing and construction sectors are relatively low, compared to emissions from other sectors. The preponderance of traditional housing, with materials that are less polluting than concrete, especially in rural areas, may explain these levels of emissions. Figure 15.3 displays a comparison of greenhouse gas emissions per capita and CO2 emissions per capita in sample countries.

The level of greenhouse gas emissions per capita is highest in South Africa (7.48 tonnes) and in Cameroon (5.19 tonnes). The population of South Africa (about

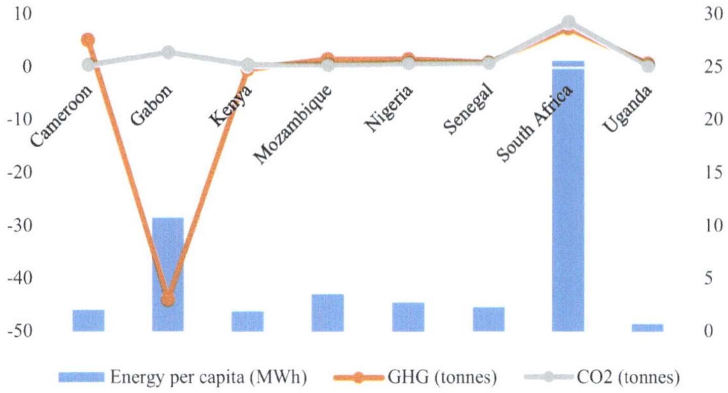

Fig. 15.4 Primary energy use, CO2, and GHG emissions per capita in 2016 (Global Change Data Lab, 2018)

57 million in 2016) was more than twice that of Cameroon. In South Africa, the emissions from the electricity and heat sector explain the high levels of both CO2 and greenhouse gas emissions per capita. In Cameroon, emissions from land-use change and forestry represent an important share of per capita figures; the sector represented 57.6% of total greenhouse gas emissions of the country. CO2 emissions per capita in Gabon (2.64 tonnes) is relatively high driven by the industry sector, but these emissions are largely compensated by the capture and storage functions of the large forestry sector that represents 110% of total greenhouse gas emissions of the country. Kenya also has overall a negative level of greenhouse gas emissions per capita (−0.27 tonnes). In Kenya, as in Gabon, the forestry sector contributes to capture a large share of greenhouse emissions (93.3 million tonnes), but this contribution is lowered by the emissions from the agriculture and transport sectors.

Figure 15.4 displays a comparison between the level of primary energy use per capita (in MWh) and the levels of CO2 and GHG emissions (CO2e) per capita for each of the sample countries.

The regularity of the GHG emissions' curve is interrupted in Gabon due to reasons mentioned in previous paragraphs. Gabon has a relatively high per capita energy use explained by its status as oil producer and a relatively low CO2 emissions per capita from the sector (0.49 tonnes). This is because Gabon also consumes a large proportion of its biomass and hydropower resources. South Africa has a relatively high energy use per capita, which is due to its available capacity for electricity and heat production, and a high level of CO2 and greenhouse gas emissions per capita due to the share of thermal energy in this production capacity.

The correlation is statistically significant between primary energy use per capita and CO2 emissions per capita with a factor of 0.99. The CO2 emissions in the study are those from the energy and cement industries, which explains the high correlation factor. The correlation is not statistically significant between the primary energy use per capita and the greenhouse gas emissions per capita. This confirms energy use is not the primary driver of greenhouse emissions in our sample countries.

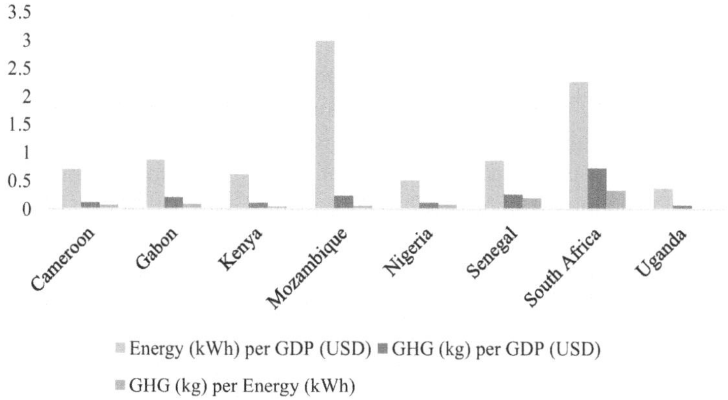

Fig. 15.5 Energy intensity and carbon intensity of countries (Global Change Data Lab, 2020a, b, c, d, e)

Figure 15.5 displays a comparison between:

- The energy intensity of the economy measured in kWh per 2011 PPP USD of GDP.
- The carbon intensity of the energy sector measured in kg of greenhouse gas emissions per kWh of primary energy use, and.
- The carbon intensity of the economy measured in kg of greenhouse gas emissions per 2011 PPP USD of GDP.

Legend: Data on energy intensity (energy use per unit GDP) and carbon intensity of the economy (GHG emissions per unit GDP) are figures of 2016. Data on the carbon intensity of the energy sector are figures of 2014, which are the most recent in the database. Figures of carbon intensity in the energy sector are relatively stable unless there is a major change in the primary energy use mix of countries, such as massive penetration of renewable energy technologies in the electricity supply mix. Therefore, we include the 2014 figures in this comparison.

South Africa is the most carbon-intensive economy of the sample, followed by Senegal (0.26 kg CO2e per unit GDP), Mozambique (0.24 kg CO2e per unit GDP), and Gabon (0.21 kg CO2e per unit GDP). With the exceptions of Gabon and Mozambique that have relatively high shares of hydropower in their electricity mix, these countries highly rely on thermal energy production powered by coal (South Africa) and diesel fuel (Senegal). Senegal still imports these fossil fuels, despite recent confirmation of oil and gas reserves, which affects both its macroeconomic and environmental indicators (energy bill and emission factor). South Africa produces coal and imports oil. The use of carbon-intensive fuels in the oil industry can explain the situation in Gabon. In Mozambique, besides hydropower units, thermal power plants installed in the country have relatively high capacities (643 MW) (USAID, 2020a, b). Despite these power capacities, access to electricity is on average 29% in the country, the majority of electricity produced being exported to South Africa. Uganda is the least carbon-intensive economy of the sample (0.07 kg CO2e per unit GDP).

Mozambique, with 2.99 kWh per unit GDP, is the most energy-intensive economy of the sample, followed by South Africa (2.27 kWh per unit GDP). The situation may be explained by the fact the database includes figures of energy production in the country and not figures of energy consumed in the economic activities of the country. Gabon (0.87 kWh per unit GDP) and Senegal (0.86 kWh per unit GDP) have relatively high carbon-intensive economies compared to other countries in the sample due to reasons explained in the previous paragraphs. Uganda (0.37 kWh per unit GDP) is the least energy-intensive economy of the sample.

South Africa (0.33 kg CO2e per kWh) and Senegal (0.2 kg CO2e per kWh) have the energy sectors with the highest carbon footprint. In both cases, the share of thermal units in the electricity production mix largely explains this situation. Kenya (0.05 kg CO2e per kWh) and Mozambique (0.06 kg CO2e per kWh) have the energy sector with the lowest ecological footprint due to the large share of hydropower and geothermal (Kenya) units in their electricity production mix.

The correlation between greenhouse gas emissions per capita and the energy intensity of the economy defined as primary energy use per unit of GDP ($R = 0.178$) is statistically not significant. The correlation between the energy use per capita and the energy intensity of the economy ($R = 0.48$) is statistically not significant. The correlation between greenhouse gas emissions per capita and the carbon intensity of the economy defined as per capita equivalent CO2 emissions per unit GDP ($R = 0.157$) is not statistically significant in sample countries.

4 Discussion of Findings

With the exception of South Africa, the levels of greenhouse gas emissions observed in sample countries are relatively low compared to other regions of the world. More concerning is the trend of emissions, which is constantly increasing, for example in South Africa, the greenhouse emissions increased by 44% between 2000 and 2016 to reach 497.39 million tonnes in 2016. Similar trends are observed in other countries of the sample during the same period: Uganda (+78%), Mozambique (+38%), Nigeria (+25%), and Senegal (+22%) (Global Change Data Lab, 2020a, b, c, d, e). Ritchie and Moser's (Ritchie & Roser, 2020) claim that the level of CO2 emissions may reach 100 gigatonnes by 2050 in the absence of strong policies to mitigate climate change. However, there have also been positive trends recorded during the period, including in Gabon, where the carbon capture capacities were multiplied by a factor of 13, and in Kenya, where the compensation of greenhouse emissions by carbon capture increased by 51%.

South Africa combines a carbon-intensive and energy-intensive economy. The greenhouse gases are emitted predominantly by the country's heat and electricity generation sector. Over 91% of South Africa's electricity production is from thermal units such as coal power plants.

Another example is that of Nigeria, which combines a relatively low carbon-intensive and low energy-intensive economy, despite the fact the country is ranked second in our sample in terms of annual greenhouse gas emissions, with 481.02 million tonnes in 2016. Two factors can explain this observation: the primary energy mix and demography. In 2016, the Nigerian primary energy mix was made of 76% biomass and biofuel energy, 14% oil, and 9% natural gas (IEA, 2020a, b, c). Other energy sources are coal, hydropower, solar photovoltaic and wind energy. Biomass energy is considered carbon neutral, which explains the relatively low levels of greenhouse gas emissions from the energy sector. With regard to the demographic factor, the population was estimated at 186 million inhabitants in 2016, with 1.6 tonnes of equivalent CO2 (CO2e) emissions per capita, which contributed to lower the absolute figure of the economy carbon intensity.

Gabon combines negative greenhouse gas emissions per capita and an economy in which carbon intensity and energy intensity are higher than in Kenya and Nigeria. Two factors can explain this observation: renewable energy systems in the electricity supply mix and demography. Gabon consumes a large proportion of its biomass and hydropower resources. Hydropower represented 51% of the 720 MW installed electricity generation capacity (Ngari, 2020). The presence of an oil industry explains the carbon intensity of the economy and the relatively high level of GDP per capita in 2016 (14,334 PPP USD 2011). These figures divided by the country population estimated at 2.01 million inhabitants in 2016, return high levels of carbon and energy intensity of the economy.

Cameroon, Mozambique, and Nigeria emit the majority of greenhouse gases (GHG) in the sector of land-use change and forestry. In Cameroon, the sector represents 57.6% of greenhouse gas emissions, and in Nigeria, 38.5% of emissions. Energy initiatives to curb the trend of GHG emissions in these countries should primarily target measures to reduce deforestation, forest degradation, and enhance carbon stocks (REDD+), through diversification of the primary energy supply. The reduction of the share of biomass energy in Cameroon and Nigeria supply mixes requires investment in alternative fuels such as biofuels (e.g pellets for cooking) from waste recycling and hydropower for electricity generation. In Mozambique, an investment in decentralized renewable energy systems and network infrastructure should contribute to increase access to electricity and lower pressure on biomass resources.

In Kenya, Senegal, and Uganda, agriculture is the primary contributor to greenhouse gas emissions, with over one-third of emissions. Therefore, energy initiatives in these countries should target mechanisms to integrate renewable energy systems such as solar photovoltaic and wind pumping and irrigation in the sector. In addition to energy-based solutions, other levers of action to mitigate CO2 emissions in the sector include sustainable agriculture practices such as crop rotation and combination of food crops with other biomass species to increase the plant cover and prevent land erosion. Aside from CO2, agriculture is a source of relatively significant emissions of methane, which is another greenhouse gas with higher radiative forcing. Therefore, these countries could also invest in organic composting and have prohibitive legislation on post-harvest burning of crop residues. In addition to investment in sustainable land management and agriculture practices, rural electrification is another indispensable lever in all sample countries, considering that the majority of rural communities in these countries still rely on biomass energy, and kerosene, especially in Nigeria. According to Chirambo (2018), by focusing on rural electrification and linking access to energy to sustainable agriculture practices, Africa could achieve sustainable development goals by 2030.

Transport is another sector that requires attention in all countries of the sample, especially in Kenya, Nigeria and South Africa, where it represents a relatively important share of greenhouse gas emissions. Affordable access to transport is necessary for the key economic activities. However, the majority of vehicles, which are predominantly fossil fuel-powered vehicles, have a significant ecological footprint. Countries in Sub-Saharan Africa have, recently and at different degrees, invested in mass transport with buses and trains. However, this level of investment is still low compared to the demand in the sector. The recent tragedy in Cameroon (2016 Eseka train derailment) provides a reminder of the contrast between the demand for mass transportation and the options accessible to the populations. In addition to the limited availability of mass transport, countries face the issue of second-hand cars that are imported in the region with higher energy demand and higher levels of greenhouse gas emissions.

5 Conclusion

The chapter provides an interpretation of the figures pertaining to study the relationships between economic growth, primary energy use, and greenhouse gas emissions. The analyses covered eight countries in the sub-Saharan Africa region, including countries with the relatively high level of greenhouse gas emissions, such as Nigeria and South Africa and countries with relatively low levels of emissions, such as Senegal and Kenya. The findings show that the macroeconomic structure and distribution of activities between the various sectors have an impact on the level of primary energy use and on the level of greenhouse gas (CO2e) emissions. Therefore, any attempt to mitigate greenhouse gas emissions or to improve energy access should build from this macroeconomic structure.

So far, the debate on the transition to energy sustainability in Sub-Saharan Africa has repeatedly clashed with these countries' quest for increased economic growth measured in terms of gross domestic product. However, the study shows that tailored actions in some key sectors that are specific to each country may return a high value in terms of transition to energy sustainability (SDG-7) and climate change mitigation (SDG-13) through reduction of greenhouse gas emissions without harming the economy. These actions could also impact SDG-11 on responsible production and consumption. For instance, in Senegal, investments in solar water pumping and irrigation may return better environmental (emissions reduction) and economic (productivity in the agriculture sector) values than promoting rooftop solar photovoltaic energy in buildings. In South Africa, recycling waste to energy in the manufacturing and construction sectors combined with investments in large hydropower systems for electricity generation is necessary to curb the growing CO2 emissions pattern.

Addressing CO2 emissions in the transport sector by investing in mass transportation, in both vehicles and fuels, is a key lever of action in all the countries in our sample. Alternative transport fuels such as jatropha oil have been experimented with in many countries of the region, including Senegal, with interesting results, especially on capacities to recover lands eroded by food crops (Dafrallah & Ackom, 2016).

These study findings should support the ambitions of countries in the Sub-Saharan Africa region to make a transition towards a sustainable energy future; ambitions stated in the post-2015 Development Agenda that include the Sustainable Development Goals. Beyond a quantitative review of the relationships between economic growth, energy use and environmental protection, the study's findings provided a reminder of the need to have a more inclusive approach to addressing the collective action problems of energy poverty, economic deprivation, and climate change.

Future studies could refine these findings by focusing on groups of countries that have similar levels of development (GDP per capita), energy use (energy per capita), or level of greenhouse gas emissions (CO2e). This could support tailor-made recommendations to address one or more dimensions of the energy-economic-environment nexus.

References

Adom, P. K., Bekoe, W., Amuakwa-Mensah, F., Mensah, J. T., & Botchway, E. (2012, November). Carbon dioxide emissions, economic growth, industrial structure, and technical efficiency: Empirical evidence from Ghana, Senegal, and Morocco on the causal dynamics. Energy, 47(1), 314–325. doi:https://doi.org/10.1016/j.energy.2012.09.025.

Chirambo, D. (2018, October). Towards the achievement of SDG 7 in sub-Saharan Africa: Creating synergies between power Africa, sustainable energy for all and climate finance in-order to achieve universal energy access before 2030. Renewable and Sustainable Energy Reviews, 94, 600–608. doi:https://doi.org/10.1016/j.rser.2018.06.025.

Dafrallah, T., & Ackom, E. (2016). Analysis of national Jatropha biodiesel programme in Senegal. *AIMS Energy, 4*(4), 589–605. https://doi.org/10.3934/energy.2016.4.589

Esso, L. J., & Keho, Y. (2016, November). Energy consumption, economic growth and carbon emissions: Cointegration and causality evidence from selected African countries. Energy, 114, 492–497. doi:https://doi.org/10.1016/j.energy.2016.08.010.

Eustachio, J. H., Caldana, A. C., Liboni, L. B., & Martinelli, D. P. (2019, December). Systemic indicator of sustainable development: Proposal and application of a framework. Journal of Cleaner Production, 241. doi:https://doi.org/10.1016/j.jclepro.2019.118383.

Global Change Data Lab. (2018). *Countries (2016): Primary energy use, GHG emissions, CO2 emissions*. UK. Retrieved November 12, 2020, from https://ourworldindata.org/

Global Change Data Lab. (2020a). *Annual GHG emissions*. UK. Retrieved November 12, 2020, from https://ourworldindata.org/grapher/

Global Change Data Lab. (2020b). *Countries (2016)*. Emissions per capita. Retrieved November 12, 2020, from https://ourworldindata.org/

Global Change Data Lab. (2020c). *Countries (2016): GHG emissions per sector*. UK. Retrieved from https://ourworldindata.org/grapher/

Global Change Data Lab. (2020d). *Countries: Carbon intensity economy (2016), energy intensity economy (2016), carbon intensity energy (2014)*. UK. Retrieved November 13, 2020, from https://ourworldindata.org/grapher/

Global Change Data Lab. (2020e). *Our world in data*. UK. Retrieved November 18, 2020, from https://ourworldindata.org/

IEA. (2020a). *Global energy review 2020*. International Energy Agency. Retrieved from https://www.iea.org/reports/global-energy-review-2020/global-energy-and-co2-emissions-in-2020

IEA. (2020b). *Nigeria Total primary energy supply (TPES)*. International Energy Agency. Retrieved from https://www.iea.org/countries/nigeria

IEA. (2020c). *Renewables 2020: Analysis and forecast to 2025*. International Energy Agency. Retrieved from https://www.iea.org/reports/renewables-2020

IPCC. (2018). *Global warming of 1.5°C. AnIPCC Special Report on the impacts of global warming of 1.5°C above pre-industrial levels and related global greenhouse gas emission pathways, in the context of strengthening the global response to the th*. Geneva: In Press. Retrieved from https://www.ipcc.ch/sr15/

Ngari, A. (2020, October 13). Tackling Gabon's future energy needs scientifically. *ESI Africa*. Retrieved November 20, 2020, from https://www.esi-africa.com/industry-sectors/generation/tackling-gabons-future-energy-needs-scientifically/

Ritchie, H., & Roser, M. (2020, August). *CO2 and greenhouse gas emissions*. Our world in data.. Retrieved November 18, 2020, from https://ourworldindata.org/co2-and-other-greenhouse-gas-emissions

USAID. (2020a). *Mozambique power Africa fact sheet*. USAID. Retrieved from https://www.usaid.gov/powerafrica/mozambique

USAID. (2020b). *Power Africa fact sheet: South Africa*. Washington D.C, US.. Retrieved October 2020, from https://www.usaid.gov/powerafrica/south-africa

Wang, X.-C., Klemeš, J. J., Dong, X., Fan, W., Xu, Z., Wang, Y., & Varbanov, P. S. (2019). Air pollution terrain nexus: A review considering energy generation and consumption. *Renewable and Sustainable Energy Reviews, 105*, 71–85. https://doi.org/10.1016/j.rser.2019.01.049

Zaman, K., & Abd-el.Moemen, M. (2017). Energy consumption, carbon dioxide emissions and economic development: Evaluating alternative and plausible environmental hypothesis for sustainable growth. *Renewable and Sustainable Energy Reviews, 74*, 1119–1130. https://doi.org/10.1016/j.rser.2017.02.072

The manufacturer's authorised representative in the EU is Springer Nature Customer Service Centre GmbH, Europaplatz 3, 69115 Heidelberg, Germany. If you have any concerns regarding our products, please contact ProductSafety@springernature.com

Printed and bound by CPI Group (UK) Ltd, Croydon, CR0 4YY
30/09/2026
02225498-0003